THE TECHNOLOGY OF MAN

Endpapers:

Applied skill is increasing precise knowledge at a
rate previously undreamed of, and in no sphere
more spectacularly than in the observation of the
Universe and speculation about its origin.
Astronomy is the oldest of all the sciences:
yet now the pace of progress is at its
keenest. The outer planet Jupiter had been
observed by Man, as the second brightest object
in the night sky, long before it could be computed
as being some 400 million miles from Earth. Less
than four centuries ago, Galileo, using the newly
invented telescope, saw more of Jupiter than any
'naked-eye' astronomer and identified four
satellite moons out of the 13 that have since been
logged. In 1664 Robert Hooke, using a better
telescope, saw what has come to be called the
Great Red Spot of Jupiter, but it was not
accurately depicted until 1831 by the amateur
Heinrich Schwabe. We now know it is 25,000 miles
long, roughly the circumference of Earth. In
1977 the unmanned spacecraft Voyager 1 and
Voyager 2 were dispatched to study the outer
planets. By the summer of 1979 they were
transmitting pictures of the Great Red Spot of
Jupiter. This photograph, taken from a distance
of three million miles, shows, in detail
unobtainable by terrestrial telescopes, the
thousand-year tornado in the giant swirl of gas
that constitutes Jupiter's atmosphere.

HOLT, RINEHART AND WINSTON, NEW YORK

THE TECHNO OF

CARLO M. CIPOLLA & DEREK BIRDSALL

LOGY MAN

A VISUAL HISTORY

The authors wish to express their special thanks to Mr. Harry J.Gray, Chairman of United Technologies Corporation, for his vision, imagination, and generosity in support of this book, without which the entire venture would have been impossible.

Many authorities and friends have given valuable help and advice. Grateful thanks are due in particular to Zöe Davenport, Brian Haresnape, David Holyday, Robert Kruszynski, Martin Lee, Helen Marshall, Dr. Simon Mitton, James Mosley, Dieter Pevsner, Lucy Sisman, and Greg Vitiello.

Last but not least to Raymond D'Argenio and Gordon Bowman of United Technologies Corporation.

Carlo Cipolla
Derek Birdsall

First published in the United States in 1980 by Holt, Rinehart and Winston, 383 Madison Avenue, New York, New York 10017.

Published simultaneously in Canada by Holt, Rinehart and Winston of Canada, Limited.

Library of Congress Cataloging in Publication Data

Birdsall, Derek.
  The technology of man.

  Includes index.
  1. Technology––History. I. Cipolla, Carlo M., joint author. II. Title.
T15.B57 1980 609 80–13921

ISBN 0 03 057792 6

First American Edition

The Auden quotation is reprinted from *Horae Canonicae*, © 1955 by W.H.Auden, which appears in *W.H.Auden Collected Poems*, edited by Edward Mendelson by permission of Random House, Inc.

Notes to the illustrations compiled by Allen Andrews

Designed and produced by Derek Birdsall

Printed in England by Penshurst Press

For Shirley, Ora & Joyce

10 9 8 7 6 5 4 3 2 1

CONTENTS

INDEX TO ILLUSTRATIONS AND SOURCES

THE SCIENTIFIC REVOLUTION

THE INDUSTRIAL REVOLUTION

To the first flaker of flints
Who forgot his dinner. (W.H.Auden)

Many anthropologists place the beginning of the human adventure at the moment when some primitive creature began to produce and use tools. If so, the earliest flaked flints, the most ancient polished stones mark the beginning of our history. All that followed can hardly be understood without reference to a growing number of increasingly complex tools. As Thomas Carlyle described him, *Man stands on a basis, at most of the flattest soled, of some half square foot insecurely enough. Three quintals are a crushing load for him; the steer of the field tosses him aloft like a waste rag. Nevertheless he can use tools. Without tools he is nothing. With tools he is all.* The role of tools and technology in shaping our destiny is so dramatically obvious, though, that it may distort our sense of the relative importance of the factors at play. In the presence of so many technical miracles, we may be blinded to the fact that it is not technology which is truly miraculous, but the mind behind it.

Technology is not the parent of human activity: it is its stepchild. Both technology and invention arise from historical circumstances of an altogether peculiar character and are part of the total human experience. When we try to explain why a given technological development occurred at a certain place and a certain time it is very easy to fall prey to facile determinism: with hindsight many a development may look simply like the natural, almost inevitable product of some need. But necessity is itself often man-made; and even where an unarguable, objective necessity has been an important ingredient, it has represented a challenge, but hardly the full explanation. We are frequently told this or that society developed labor-saving machinery *because* it faced shortages of labor; or that the mechanical clock was invented *because* the sun did not always shine on sundials, and in winter the water in water clocks turned into ice. In fact, when faced by a shortage of labor, one society may fatalistically accept the consequences; another may launch raids to capture slaves; another may develop labor-saving machinery. The response depends largely on the prevailing culture. Clouds and cold weather have existed always and nearly everywhere; the reason why the clock was invented in medieval Europe can be explained only in human terms.

In fact when we speak of the challenges of a given environment or the needs of a given society, we must beware of thinking in terms of purely objective, exogenous factors. The challenge counts insofar as it is perceived by man, and what matters is how it is perceived.

Actually, the characteristics of a culture not only condition the appearance of tools but also their use, destination, and spread. The microscope was invented in the 17th century; but it was another two centuries before microbiology was born. Until the development of a new philosophy and new methods of systematic empirical inquiry, the doctors who looked through microscopes at microbes could not make sense of what they saw. In the 17th and 18th centuries, when Europeans took clocks and scientific instruments to China to impress the Imperial Court and the Mandarinate, the Chinese were eager to acquire such pieces, but mostly as toys.

To say the development and spread of technology can only be explained in human terms does not, however, imply that technology is a neutral stepchild of human activity. Technology deeply affects the material culture of a society, the size and composition of its population, the composition of the labor force and its work patterns, and the physical environment. Most important of all, it affects human minds. A boy playing with a mechanical toy and a scientist using a computer will both be deeply affected in the workings of their minds, their inclinations and their curiosities by the gadgetry they are using. The technologies nurtured by a culture may easily have a cumulative effect on it.

A catalog of tools and gadgets is not a history of technology, and conversely a pictorial history of technology inevitably suffers from the fact that it must concentrate on gadgetry *per se*. This introduction and the short essays throughout the book are presented as necessary, though inadequate, correctives. They are invitations to ponder the subtle interrelation between man and his technologies – and to remind the reader that behind any tool or contraption there have been men with their own values, wants, beliefs, and peculiar ways of perceiving the relationship between themselves and their environment.

Carlo M.Cipolla

The Earth is roughly 5,000 million years old. Man has existed on it for the last 2½ million years, and that is the range of the technology of man. The archaeologist defines man by the first appearance of stone tools, the earliest of which, from level C of the Omo valley in Ethiopia, are dated at 2.51 million years old. The physical anthropologist defines man by the first appearance in hominid fossil bones of certain features based mainly on tool-using and tool-making: a reduction in the size of jaws and teeth after it was no longer necessary to tear raw flesh from a carcass; an increase in the size and complexity of the brain, with a change in the aspect of the face, coincident with better coordination of hand and eye, flexible social and behavioral activity, and vocal communication. (Skeletal and muscular modifications permitting erect posture, with changes in hands and arms for the effective use of tools, are significant but not computed, since the ancestors of man could intermittently walk erect and use random, unmanufactured tools.) Using these anthropological standards, and the dating of lava *under* which the fossil was found, Phillip Tobias, professor of anatomy at the University of The Witwatersrand announced in 1977 that he dated the skull of definitive Man (*Homo habilis*) discovered in 1972 by Richard Leakey at Koobi Fora, East Africa, and known as ER 1470, at 2.41 million years. The anthropological and archaeological verdicts on the age of man thus show remarkable consistency.

The pebble tool from Bed I of Olduvai Gorge, East Africa, depicted in its actual size (as are all the tools in this section), is 1.8 million years old. It is a multipurpose chopper, used for cutting and pounding vegetables, cutting raw flesh from a carcass, cutting up flesh for food, and cracking bones for marrow. (Cooking was an art which would emerge only a million years in the future.) The tool is still comfortable to hold in the hand, but most effective in the hand of a 10-year-old child of today. The hairy hominid who wielded it achieved a height in maturity of some 130 centimeters, 4 feet 3 inches. Although this specimen is of simple design, fashioned from a green volcanic basalt water-worn pebble by only three skillful blows from another stone, not all the tools from this site and date are crude. Some of the 11 different types of implement are so sophisticated that they have been paralleled only by tools a million years more recent. John Pfeiffer wryly declared *the tools come a million years too soon, and it is almost as if one had opened up a musty vault in the Great Pyramid of Egypt and found vacuum cleaners and television sets.*

Amid a vast number of broken animal bones found at the Olduvai Gorge, East Africa, though the majority have been classified as food debris, some show evidence of having been deliberately worked on so that they were modified to become tools. Among them, and dated at 1.8 million years, is the highly practical distal end of the shinbone of an animal of the genus *Equus*. (The "modern" horse, zebra, and donkey evolved some 3.5 million years ago, but their ancestors had existed for 10 times as long.) The upper end of this bone has been chipped by a stone tool to form sharp points which enabled the bone to be used as a pick; the lower, articular end was the joint of the bone and was of a convenient shape to be held in the fist.

The discoid scraper found at Koobi Fora, East Africa, dated at 1.8 million years, was recovered from the site of a hippopotamus-butchering area and was used for scraping the flesh off skin and bone. It cannot be presumed that these large animals were methodically killed by early man as hunted prey. Killing such a beast, unless it was helpless through age, disease, or accident, was perhaps beyond his skill. At that time, slaughter was confined to tortoises, snakes, small pigs, and similar manageable creatures. When larger animals were fortuitously dispatched, parts of their bodies were chopped into portable sections and carried to an established site for more meticulous butchering. The tools used were not only of stone. Evidence at one site indicates that the broken bones of slaughtered animals were used to cut the hide of other carcasses. However, the discoid scraper was and is an efficient tool. Using it, Louis Leakey skinned and butchered an eland within 20 minutes.

The red sandstone hand axe found in Bed II of the Olduvai Gorge, East Africa, (that is, in the stratum above the bottom 6-meter belt of the 100-meter deep gorge) is dated between 1.2 and 1.1 million years. Hand axes, tools with two faces that form an edge, are known from 1.4 million years ago, but the earliest have not been found at Olduvai. They continued until 40,000 years ago, becoming progressively thinner, straighter-edged and smaller. It is thought that this progression coincides with the development of more neural tissue in the motor area of the brain of man, allowing better coordination between the thumb and the forefinger, with a consequent increase in skill that permitted the shaping of a finer edge to the tool. The implement was used for butchering animals and preparing vegetable food. Though the early hand axe is primitive, its construction required considerably more time and effort than the making of the first pebble tools. The side view shows that an appreciable number of flake scars have been removed to form the edge.

Fire was controlled by man more than half a million years ago. Hearths have been identified in a cave at Escale, near Nice in France, and charred bones at the site have been dated at 600,000 years old. The natural sources of fire – volcanic eruptions, lightning strikes, and the spontaneous combustion of hydrocarbon outcrops – were appropriated, controlled, and stored in hearths which were banked for perpetual use. Since the evidence would have been immediately consumed, understandably none exists that primitive man rubbed sticks together; and the "flint-and-steel" method cannot be dated before 15000 BC, concurrent with the earliest known module of iron pyrites, clearly marked by the continuous flint scratchings of a fire raiser. Fire was used on a large scale in hunting, to drive animals into an ambush. The remains of 64 elephants have been found at Torralba in Spain, ringed by carboniferous evidence of 10 successive marsh fires. Domestic fire, maintained in caves, was a necessity for human survival when a glacial period of the Ice Age affected the Northern Hemisphere as far south as the Mediterranean from about 600,000 years ago. Significantly, this is the estimated date of the Escale hearths; fire has not been established as a feature of human life in Africa before the year 100,000.

The Ice Age drove men into the shelter of caves – where they fought with animals for possession – but once man had established a domicile it was fire that kept predators at bay. Socially, fire changed human life by extending the active day and allowing communal decision-taking, hunt-planning, exhortation and ritual and bardic recitation as well as merry-making. Predictable accidents – food or bones falling into the fire – led to the utilization of fire for cooking and tool-making. These were both survival techniques. Cooked meat was more nutritious (and, naturally, more tender, which allowed man's molar teeth to become smaller). Fire hardens bone, antler, or wood, making it a better tool or weapon. The small charred meat bone and the fire-hardened antler tip, both 500,000 years old, were found in China in the cave at Choukoutien where the remains of 40 individual specimens of "Pekin Man" were discovered, and where a gigantic hearth with its carbonized remains plunging 6 meters deep gave graphic testimony to the "eternal flame" of a formerly continuous fire. The antler tool was used as a finishing hammer to perfect the cutting edges of stone tools.

The elephants at Torralba, stampeded by fire 400,000 years ago, were killed by fire-hardened wooden spears of which only fragments remain. With one exception, the oldest wooden tool in the world is the spearhead (also potentially a digging tool) from Clacton, England, dated at 250,000. This 15-inch yew weapon was recovered in a marshy area where it had been impregnated with preservative minerals. Nearby were found the bones of hippopotamus and rhinoceros, but none of them showed any sign of butchery.

Where butchery was done the animals were dismembered with hand axes and cleavers, the latter being a heavy-duty tool apparently developed for cutting meat. A cleaver of the date of the Torralba site, but found in Southern Africa, shows the precision with which man, after over a million years of toolmaking, could fashion a practical tool with an effective long (5 centimeters) straight edge presenting a cutting angle of extremely high efficiency. The angled blade was indeed later to be the tool selected by Dr Guillotin when, after field tests on dead bodies, he redesigned a historically older decapitation knife to cut forcefully with an oblique edge.

The fine hand axe from Swanscombe, England, dated at 220,000, was found only 3 meters from the skull bone of a representative of "Swanscombe Man," one of the earliest specimens of *Homo sapiens*, a bud of the final flowering of the human pedigree from *Homo habilis* and *Homo erectus* to the species which includes modern Man. Swanscombe Man, who lived in an England where the recession of the glaciers permitted elephant and rhinoceros to range the Thames Valley forests, manufactured and used a strikingly high proportion of sophisticated straight-edged hand axes. The stone was first struck to detach a large flake, leaving a rough edge. A soft hammer of wood, antler, or bone had to be used to produce a very sharp and straight edge traveling all the way round the axe. The number of chipping blows required to produce the fine hand axe has been calculated as twice the requirement for the older, rougher hand axe.

The period of human development from the years 200,000 to 40,000 is marked by a spread of population into new habitats, by the shadowy beginnings of leisure and art, and by a far more definite technological and ultimately artistic revolution in the working of stone. The *Levallois technique* was invented by hand axe makers. It appeared in Africa, Europe, and Asia some 200,000 years ago but it is called after the Levallois-Perret suburb of Paris where a key site was identified. The technique demanded intellectual projection. Previously, a visible stone, suitably adapted, was converted into a tool roughly identifiable in shape with the original. Now, the final object could not be visually adumbrated. It was something that was to be made out of the invisible core of a stone which might well be three times the mass of the final tool. The tool had to be envisaged *within* the original nodule. The outline could vary; it was not essential to keep to the conventional pear-shape of the old hand axe. An estimate of the tool's thickness and of the line of its spine had to be made. The nodule was chipped by preliminary flaking. A flat platform was chopped at the base of the tool. For the tool shown 111 shaping blows were required, compared with some 25 blows necessary for older-type hand axes. Then one crucial blow was given to the platform end with maximum force and calculation. Only an expert toolmaker who knew how flint fractures would achieve the tough, very sharp instrument with a preformed backing that resulted, springing from *inside* the block of stone.

The operation demanded craftsmanship of a high order. It foreshadowed the far more intellectually complex leap into art, the sculpture of a represented notion rather than a tool. The line of succession is direct. If Aristotle's prerequisite for a work of art is accepted, that *the conception of the result exists before its realization in the material*, man the unique technologist fathered man the artist. The rhomboidal Levallois flake together with the Levallois point come from Crayford, England. Both are some 150,000 years old. The profile views show the shaping of the humped spine. The concave surface leading to the sharp edge was struck off from the bottom platform in the final decisive blow. The pointed tool is also hafted, so that it could be fastened as a spearhead to a stake; a chipped groove allowed the head to be bound on with strips of hide, sinew, or the hair of the mammoth, rhinoceros, or horse which were hunted at that time. This was the second breakthrough of the Levallois technique. From a lump of flint, a man had made a part – the spear point – and by binding it to another manufactured part – the straight shaft – he had made a new whole.

A fresh development of the Levallois technique was the Mousterian technique, named after a site at Le Moustier in the Dordogne area of southern France. By efficient and exhaustive flaking of the stone core, a maximum number of usable flakes were produced. "Retouch flaking" often produced varieties, including the denticulate scraper and the side scraper shown, both from Le Moustier and dated at 40,000 years old. The wear patterns of these tools have been closely studied.

The denticulate may have been used for working wood. Side scrapers were used in some cases for the working of skins, which suggests the use of skins for clothing. Over 60 different types of tool have been recognized as made by the new methods. Hand axe manufacture could produce a maximum cutting edge of about 8 inches. Using the Levallois technique a craftsman could produce five times the output of cutting edges from the same amount of stone.

The Venus from Brassempouy, France, 22,000 years old, is a version of fertility-goddess figurines which more often emphasized the breasts, belly, and buttocks. It is carved from mammoth ivory, a medium which could not be worked unless it had previously been heated. Though sandstone may have been used for polishing, the figure was mainly carved with an engraving tool called a burin, the most important individual stone tool of that age, since it made other tools: the bone hammer, the needle, and the harpoon.

The burin is a composite tool, being an incisor at one end and an endscraper at the bottom, which is used for honing down other tools of bone, antler, and wood. The burin was essential to make the blade tools, the flint knife, and the backed blade (a blade having a blunt spine like a penknife, and in this instance clearly hafted for attachment to a shaft), which are characteristic of the era. In the preceding flake stage of tool development, 98 per cent of the tools which have been recovered are of stone. In the succeeding blade stage 50 per cent are of bone or antler, and many tools, now decomposed, must be presumed to have been made of wood. Making blades was a far more complicated procedure than making flakes. A different kind of fracture had to be given to the original nodule, and this could not be done by blows from a stone hammer. A bone hammer had to be employed to give up to 250 strokes on a prism or near-cylindrical nodule. These peeled away the outside, but in the process they produced tools with spines already formed by chipping, so that a mass of stone would produce five times the number of cutting edges when worked on by blade-tool than by flake-tool technique.

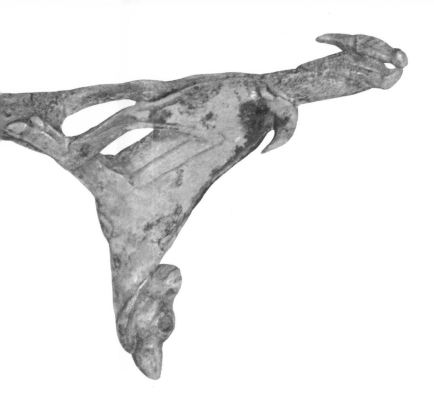

Spear throwers, made in antler, bone, and wood, were hand-held and ideally extended the exact length of a man's arm. They have been found in Europe, dating from 15000 BC, in North and Central America, and in Australia, where the weapon is called the *woomera* but has not been dated before 10,000. Spear throwers extend the range of a missile by about 75 per cent; present-day tests have increased the flight of a spear from 40 to 70 meters. Using a spear thrower, an Eskimo hunting reindeer could overpass animals at the edge of a grazing herd, which were specially posted for their alertness as scouts, and fell a more distant animal. When accidents occur in the middle of a herd, reindeer graze on while the victim dies. By avoiding an incident on the fringe which would panic the herd, the hunter had time to select and attack other animals. The spear thrower shown, 15,000 years old, has a counterweight essential for the balance of the weapon.

The bone needle, which sewed sinew or hair threaded through its eye, is from 15000 BC and was made with a burin, as was the single barbed harpoon and the harpoon with the bevelled wedge which was inserted into the haft of the shaft. The double-barbed spear point of antler, from 15000 BC and from southwest France, has two additional "bayonet" protuberances above the wedge to fix it more securely.

The spear thrower relied on an adaptation of the human body to exploit the mechanical principle of the lever. The bow stored energy which had been created by the action of human muscle on vegetable sinew, and suddenly released it to achieve a new mechanical potential. The arrowheads which made the projectile effective were at first the specialty of a culture called the Aterian (from a site at Bir-el-Ater in Tunis) which ranged across North Africa from eastern Libya to the Atlantic. They applied Levallois and Mousterian skills particularly to the bifacial retouch working of warheads for spears and arrows. This Aterian arrowhead is dated at 15,000 years and is typical of many found in North Africa of that age. With the exception of some found in Germany no arrowheads recovered in Europe date earlier than 8000 BC.

Materials, sources, and dates of the arrowheads mounted are, in descending order: top two, obsidian, East Africa, 8000–6000 BC; third, flint, North Africa (Sudan), about 4000 BC; fourth, flint, Europe (Ireland), about 3000 BC; bottom three, quartz, North America, about 1000 BC.

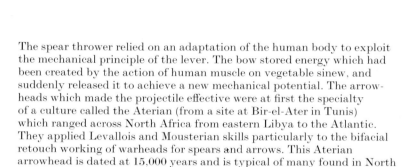

Cave painting, an art largely concentrated in the Dordogne area of France, the French Pyrenees, and the Cantabrian mountains in Spain, were mainly made between the years 15000 and 10000 BC. Often executed in almost impenetrable sites, they reflect a magical or ritual significance to the hunting of game which then preoccupied men's minds. But in a resurgence of cave painting localized in eastern Spain and continuing until 5,000 years ago, human figures – rare in the older work – are rendered more frequently and there is a far more skillful portrayal of action and movement. The stag hunt, originally painted in red and black ochre and possibly 10,000 years old, shows bows and arrows in use.

The Pesse log boat, some 4 metres long, was found in northern Holland and is dated at 6400 BC. The date of the earliest boat is difficult to fix. There is no evidence of human habitation in Crete before 9000 BC and domestic animals do not appear until 7000 BC. Mallorca, Minorca, and Corsica were not occupied until about 7000 BC. But it is known that Australia was populated some 50,000 years ago; as the continent was never part of the southeast Asian land mass, man presumably reached it by some sort of boat. If so, a relic may yet be recovered.

THE NEOLITHIC REVOLUTION

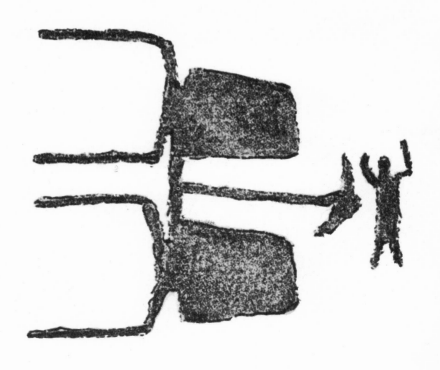

Overleaf: rock engraving of scratch plough

In the Middle East, soon after 10000 BC, small groups of people first succeeded in domesticating some types of animals and growing some kinds of grains. Similar experiments with different types of plants apparently had been under way thousands of years earlier in Southeast Asia. Independently of these developments, after 7000 BC, small groups of people on the American continent began to experiment with primitive forms of food production.

These humble beginnings paved the way to a dramatic world-wide change. Slowly but irresistibly, more by the movement of people than by the transmission of know-how, the new modes of production were broadcast all over the globe; and wherever they made their inroads, savagery gave way to civilization. Man, previously a hunter and a gatherer of wild fruits, turned farmer and shepherd. The temporary camps of nomads gave way to villages and towns. Increasingly complex and larger societies flourished in places previously occupied by small nomadic bands of hunters. The Neolithic Revolution was a watershed. Paleolithic man left behind more bones and fractured skulls than artifacts – and even his artifacts were mostly the weapons of a nasty, brutish hunter. With the advent of the agricultural age, the richness and variety of man's tools and manufactures increased spectacularly.

From a purely technological and economic point of view the sequence of events looks clear. But in human terms the picture is puzzling. How does one explain the seemingly independent discovery of agriculture and husbandry in widely separate and vastly different parts of the globe? Changes in climate and ecology contributed to the Neolithic Revolution. But what undoubtedly was a necessary precondition was certainly not a sufficient cause. There must have been changes in the levels of culture corresponding to the first successful attempts to domesticate plants and animals: more specifically, there must have been subtle yet substantial changes in the attitude of men to animals and plants.

The illustrations that follow may help to recapture the dramatic leap forward in the level of economic and technological efficiency made possible by the Neolithic Revolution. But the cultural changes which were both its precondition and a constituent of it remain obscure and impossible to illustrate.

The sickle blade, dated at about 10000 BC, is made from flint. It was used for cutting cereal, which is established by the presence on the blade of layers of silicon dioxide. Silicon was secreted in the outer layers of some cereal stalks as a defense mechanism against the plant's being chewed by animals, some of which have an aversion to silicon. This element was deposited on the blades of the sickles with which men cut the stalks, later becoming oxidized into an identifiable gloss.

The sickle handle, made of antler or wild goat's horn, is grooved to hold blades made from the volcanic glass, obsidian, the blades being held in the groove with bitumen. The sickle, from Mount Carmel, is dated at 8000 BC or earlier.

The Ice Age ended in about 8000 BC. It was followed by a drying out of the environment in the Middle East. Until recently, it was thought that men and animals, desperate with thirst, came together at oases, and the men tamed the animals because the beasts could not survive if they ran away. It was then deduced that the domestication of animals coincided with organized agriculture, of which no traces had been found dated earlier than 5000 BC. But during detailed excavation and study of caves on Mount Carmel, clear indications were traced of more ancient settlements, from the early hand-axe cultures onwards. Other relics accompanying the tools in these layers of civilization established that domesticated grains were growing in that area in about 18000 BC, long before the Neolithic Age which was supposed to have ushered in agriculture.

It is now recognized that sheep and goats and barley were indigenous to high altitudes. The people of 18000 BC were not sedentary (as systematic agriculture would require) but they exploited different zones of the Levant, driving their sheep to higher levels in summer and lower in winter, and using all the varieties of wheat and barley that grew at different levels. Hunting of larger game, such as the gazelle and the Palestinian lion and wolf, diminished and the goat was introduced around 10000 BC. Systematic agriculture then began to be adopted, with a consequent concentration of population: cultivation of grain allows a high return from a small area of land. The next step was irreversible. People invaded the lower valleys, and towns began to be formed. Jericho, the earliest city in the world, was first inhabited before 8000 BC, the date when an enterprising leader ordered the erection of a wall 4 meters high with towers twice that height to protect not just the inhabitants, but the springs of water on which their subsistence depended and the reserve of grain which they stored.

The walls of Jericho show distinct layers dated at 8000, 7000, and 6000 BC. By the latter date the site had become a habitation of up to 3,000 people. Such an accumulation could not live solely by hunting and local food gathering. Trading was an unlikely possibility for their survival, since they had only water and bitumen and salt to trade. The reason for the community's survival was the organization of agriculture in a methodical way. Populous Jericho, which finally had an area of 11 hectares, 27 acres, had to undertake social organization – providing for defense and for storing food, and leading to the recognition of craftsmen and specialists – and the inhabitants developed religious organization, notably in the ritual burial and decoration of the dead. The Neolithic Age proved revolutionary in the development of agriculture and the parallel concentration of communities.

The saddle quern, or mortar, from Jericho, with its pestle for grinding down wheat, dates from 6–7000 BC. (The wheat grains shown are modern.)

Firing – the fusion of less stable substances into a more durable form –
has been traced back 25,000 years on evidence from the site at Dolní
Vestoniče, Czechoslovakia, of a kiln where a mixture of clay and bone
was fired into a very hard material. They had the technique for making
pottery, but as yet there was no demand. The earliest known pottery
comes from Japan and is dated 10500 BC. Fragile pottery from wet clay
that was dried in the sun began to be made in Mesopotamia around
7500 BC. Efficient baking, and the introduction of the slow potter's
wheel, turned by hand, were features in Mesopotamia by about
5000 BC. The fast wheel, turned by a mechanism, did not appear until
about 1500 BC. The bowls from Arpachiya in northern Iraq date from
about 5000 BC. Gradual improvements in kiln manufacture introduced
the possibility of higher temperatures and more effective pottery: modern
analysis has shown that pottery made in 3000 BC was baked at a
temperature of 900°C. It required a comparatively minor increase in
technological skill to build kilns that would give a heat of 1050°C, the
temperature necessary to melt copper. In this fashion kiln pottery would
aid the development of metallurgy.

The oldest textile known is the wool cloth recovered from Çatal Hüyük in central Turkey, a city almost as old as Jericho and three times as large. It dates from 6000 BC. Spinning and weaving were then only of recent development; no older looms have been recovered. The spindle whorl, which drew down the woolen yarn after spinning, was found in Switzerland and is dated 3000 BC.

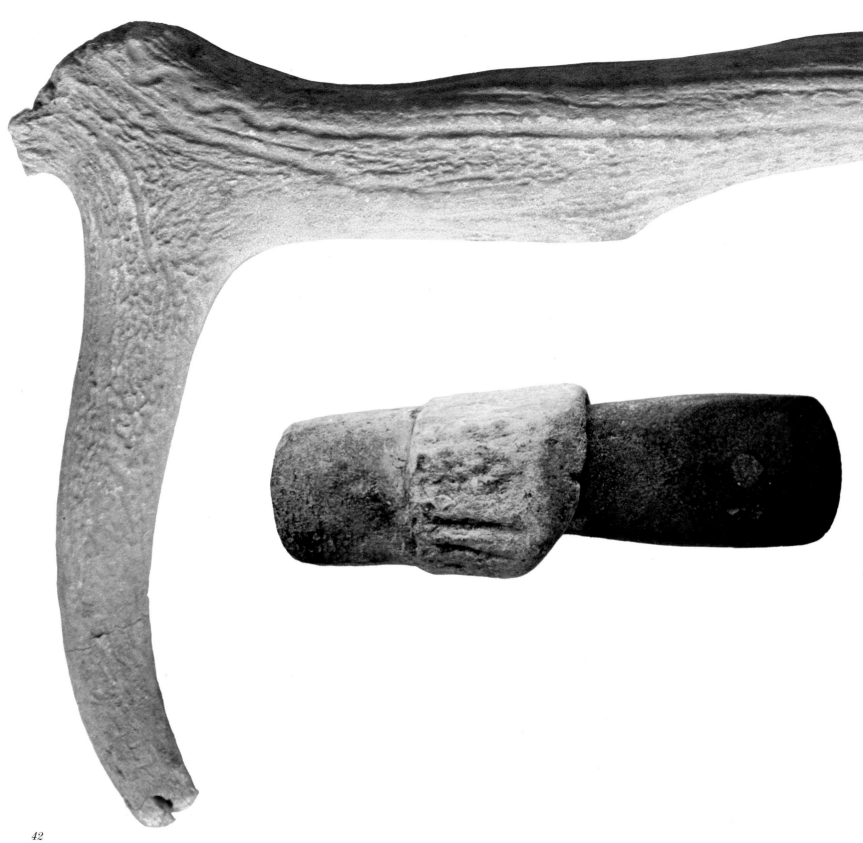

42

The antler socket holding the stone adze characterizes a further stage of tool development, designed to ease the craftsman's effort. When the adze was driven into wood by a blow from a stone hammer, the sleeve absorbed some of the shock, giving less vibration; it also enabled the tool to be manipulated and eased out of the wood more easily. The composite tool was found in Switzerland and is about 5,000 years old. Adzes were used for chopping trees and timber and for shaping the beams of the long houses of the time, which were made entirely of wood.

The miner's pick made of antler, originating between 2000 and 1800 BC, was found in Grimes Graves, Norfolk, England, in a flint mine. A sophisticated pit with conventional shafts and galleries, the site was used as an industrial repository where blanks of flint tools were stored for future orders and for export trading.

Long before there was literature there was writing. But it was a passionless, utilitarian craft, using design as what Professor E.A.Speiser called *the incidental by-product of a strong sense of private property*, claiming no vision of that extension of personality which Man had already expressed in uninhibited painting. The 20,000-year-old paintings in the Altamira cave were most probably set down as a permanent altarpiece fronting religious rites to invoke the god of the chase; they convey the emotion aroused by a bison hunt which kills both beasts and men as well as recording the technique of slaughter for food. Painting on the new medium of manufactured pottery began as a craft in Mesopotamia in about 5000 BC; and as Professor C.D.Darlington wrote: *The painter was later to become a writer, and his varying designs projected the use of symbols from a magical past into a practical and literate future.*

The first steps in literacy were taken to systematize taxation. The earliest surviving samples of writing are agricultural production returns. In Sumeria, the biblical land of Ur in southern Mesopotamia, prosperity was thriving in what was the most advanced civilization in the world. Prosperity was based on property, and its fruits in investment, tribute, rents, and commerce. A proportion of these fruits were destined for the temples, where the priests demanded strict profitability on behalf of their gods, and therefore insisted on intricate accounting. By about 3600 BC a system of recording the details of production was devised. First, pictorial representations of products were used as mnemonics, and symbols were invented to represent their number or quantity. Much later this pictography, expressing objects in visual pictures, was extended into ideographic writing which could also represent an associated idea: a picture of the Sun could imply "heat." At an early stage the system became so complicated that it was necessary to create a class of scribes who simplified the written symbols into shorthand outlines which could more easily be drawn in wet clay, which was then dried as a tablet of record.

The fact that the earliest specimens of man's writing are inscriptions of glebeland harvesting justifies Professor J.D.Bernal's assessment of writing, that this, the *greatest of human manual-intellectual inventions, gradually emerges from accountancy.* Yet it broadened swiftly. Within two or three centuries of the first known Sumerian pictographs, the priests of Egypt had developed pictograph *hieroglyphics* (the Greek-based word means sacred carvings). These were primarily concerned with the fame of the divine Pharaohs and the celestial (and occasionally earthy) aspirations of human desire. Egyptian graphic representations may well be considered as an expression of art as well as a breakthrough in technology, making writing the younger brother of painting.

Early Egyptian hieroglyphs dating from about 3000 BC have been found inscribed on wood, sometimes very finely sculptured in high relief, and afterwards painted. Simultaneously in Mesopotamia the pictographic forms diminished in number. From the 2,000 pictograms identified in tablets of 3600 BC, the number of symbols shrank to only 600 recognized in relics of 2900 BC. The inscription on a polished stone component of the "Blau Monument" found in Sumeria in the Tigris-Euphrates basin is a fine specimen of the later period. The five-column Mesopotamian tablet dated about 2800 BC gives an accounting of fields and crops.

The demands of speedy, efficient writing forced the formalization of the pictograms into cuneiform (i.e., wedge-shaped) characters inscribed with a stylus made of reed or wood and shaped like the wedge of a chisel blade, the cutting edge of which was periodically restored with a "pencil sharpener" of pumice stone. Both pictography and its descendants, cuneiform and hieratic script, represented universal objects and related ideas; they could, therefore, with great ingenuity be read by strangers who did not know the language in which the scribes were thinking. The next stage in writing was the alphabet – the representation of sound by written symbols or letters. Though achieving complex fluency, this artifact created a literary Babel dominated by differences in individual languages.

Even accepting that writing began as a record of what was tedious to remember, and had no association with creative art, we should not discount the importance of spoken lore. Certainly at one time illiteracy was a mark not only of aristocracy but of artistry, and literacy was the concern of docile clerks. What was remembered was well remembered, well spoken in its recall, and at the last only grudgingly remitted to writing. Homer, like Winston Churchill, recorded only the drama of passionate living. The rest was left to research assistants, or the more meticulous historians.

The most civilized sector of the Old World, the area bounded by the Nile and the Euphrates, advanced from the Stone Age in about 3750 BC to launch within two or three centuries not only the utilities of mining, metallurgy, and architecture but the fundamental innovations of writing, the wheel, and the boat. Since boatbuilding is prehistoric, its progress can only be inferred by studying primitive peoples who can still be observed. It begins with naturally floating objects like dead wood or dead animals, inflated by the gases of corruption. It proceeds to the gathering of wood or reeds to shape into a boat, or the inflation of animal skins to make a raft. Paddles were a logical development from the movement of the paddling hand. Oars were more sophisticated since they demanded adaptation of the principle of the lever. Sails, which are not a notably simple mechanical device, had an easy start in Egypt, where the prevailing wind blows straight up the Nile and the current naturally flows down; by and large, using sail or drift, one could set an acceptable course, and this was a comparatively early discovery.

In Mesopotamia, rafts of inflated skin were fitted with a timber deck and used to carry cargo. Circular reed coracles were rowed as lighters. The advantage enjoyed by Babylonia was that it was naturally rich in the petroleum product bitumen, which in certain locations seeped from the ground. It was widely sold for export but had valuable domestic use. It was melted and mixed with sand, lime, and chopped reed fiber to make an asphalt mastic which had two principal uses: to waterproof the floors and walls of bathrooms and water closets, and to caulk boats and ships inside and outside – all this in the 3rd millennium BC. The bitumen-caulked coracle, called in Arabic the *quffa*, is seen on the Tigris today, apparently of the same size as is shown in a bas-relief from Nineveh of the 7th century BC.

Reed boats, made out of bundles of bulrushes, were used in Mesopotamia and in Egypt, even for sea-going purposes, and delicate models of these vessels were placed in tombs. When wood was later used to make a more durable hull, the design was still reminiscent of the reed construction, the mast stepped well forward, the steering effected by two long paddles lashed to the side.

46

The first wheels appeared on vehicles at about the same time as the potter's wheel, which was first seen in Sumer about 3500–3250 BC. But wheeled vehicles were not dependent on the potter's wheel for their adoption. They were accepted for their own virtue rather than after a theoretical study of rotary motion. Wheeled vehicles, for instance, reached Egypt in 1600 BC, some 1,150 years *after* the potter's wheel, and they reached Britain in about 500 BC, 450 years *before* the introduction of the potter's wheel to southern England and 900 years before its entry into Scotland. Wheeled vehicles were preceded by the sled, which was in use by 5000 BC, and even in backward Britain the sled was transporting monoliths to Stonehenge by 2000 BC. The earliest graphic likeness of a cart is, in fact, an outline of a sled equipped with wheels instead of runners.

Three-piece plank wheels are the earliest form known, a fact which contradicts the theory that wheels derive from tree-trunk rollers placed under sleds for arduous transport. About 2500 BC, two-piece plank wheels were depicted in mosaic panels of shell and colored stone found in the cemetery of the kings of Ur and known as "the royal standard of Ur." The onagers, or wild asses, harnessed four abreast, are drawing a ceremonial wagon in the uppermost sequence which shows the king reviewing prisoners of war. But they are harnessed to lighter war cars in the chariot charge depicted at the bottom. Onagers were succeeded by horses, and plank wheels by neat spokes and felloes. Yet even the draft sled continued in occasional ceremonial use in Ur. Sleds were used there for royal funerals a thousand years after wheels became common.

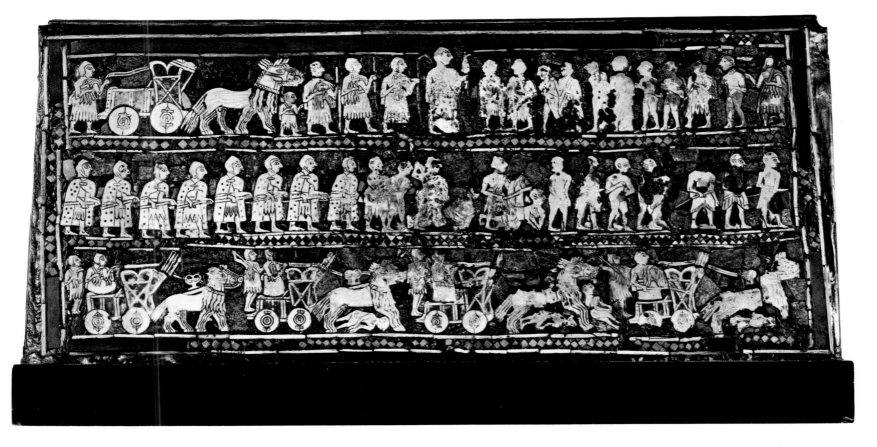

The earliest illustrations of a scratch plow drawn by oxen occur on engraved cylinder seals of the Sumerian civilization in Babylon dated about 3000 BC. Similar implements were not to reach central and western Europe for nearly 2000 years. The earliest plow was merely an animal-powered version of the wooden hoe which was scraped across the ground, or of the digging stick with which, long previously, women grubbed for roots. Wooden models from Egyptian tombs, placed there to provide a future means of subsistence for the dead, illustrate the use of the plow in Egypt by 1900 BC. The common introduction of an iron blade to the plow was delayed until the collapse of the Hittite empire in the 12th century BC, for the Hittites had kept their expertise in iron technology as exclusive as possible. Very soon, iron was used in Egypt for plowshares and sickle blades; previously sickles had had curved lengths of serrated flint set in a haft of wood or bone. A papyrus found in an Egyptian tomb of about 1000 BC shows scenes of a priestess plowing and reaping with a sickle. It is a reminder that agriculture, which began as women's work, was a seasonal specialization which at times eased or released its labor force (as year-round hunting never did) for participation in crafts such as textile working and pottery, or vocations such as the priesthood.

Lifting water has been an ergonomically awkward necessity since the utilization of the first well and the inception of fluvial irrigation. It was always handier to pull *down* on a rope which was lifting a bucket *upwards*, but the pulley wheel which permitted this simple transference was slow to be developed. The first mechanical water hoist was the bucket mounted on a counterweighted swiveling pole known by its Arabic name of *shaduf*. (In English it was sometimes called the *swipe*, a variant of *sweep*.) The shaduf was illustrated on a cylinder seal of the Akkadian period of Babylon dated around 2300 BC and was not represented in Egyptian graphic art until seven centuries later. A remarkable innovation in its time, permitting one man to raise 600 gallons of water a day, it is still in use, barely changed in design or efficiency.

Pottery was practiced in southwest and southeast Asia before 6000 BC; but the hardening by fire of shapes made from clay was apparently common to every civilization. The earliest identified part of a potter's wheel was found at Ur, and dated around 3250 BC. A complete clay disk of 90 centimeters diameter has been found in a potter's tomb at Erech in Mesopotamia, dated about 2000 BC. A series of Egyptian tomb pictures extending over some centuries from 2500 BC shows every aspect of pottery making. Decoration of pottery by shaping of the clay, coloring, and the use of different clays for different colors and varying temperatures, was practiced long before the adoption of the potter's wheel.

Basketry and matting presumably were developed before textiles, because it is conjectured that man contrived the simpler craft of plaiting before learning to twist fibers into a thread and use the threads for the much more complicated basketry we call weaving. But there is no conclusive evidence of this; the oldest existing textiles are as ancient as the oldest existing basketry, dated around 5000 BC and found in Egypt. Rope making had been perfected many thousands of years earlier, and had produced the net, invaluable not only for fishing but for carrying burdens. Spinning in Egypt for the manufacture of linen was almost entirely of flax fiber; silk and cotton were unknown, and wool was considered unclean. Spinning began with the twisting of prepared fibers between the hands, or between one hand and the cheek or thigh.
The spun thread was wound onto a stick which developed into the spindle, containing the finished thread through constriction in its collar. The standing woman depicted in an Egyptian tomb of 1900 BC uses the weight of the rotating spindle to twist the unspun threads into fine yarn. The model of a combined spinning and weaving room found in a tomb at Thebes, dated probably a century or more later, shows a horizontal ground loom. Ancient Egyptian linen shows thread finer than anything fashioned on modern machinery, sometimes accommodating 540 warp threads to the inch.

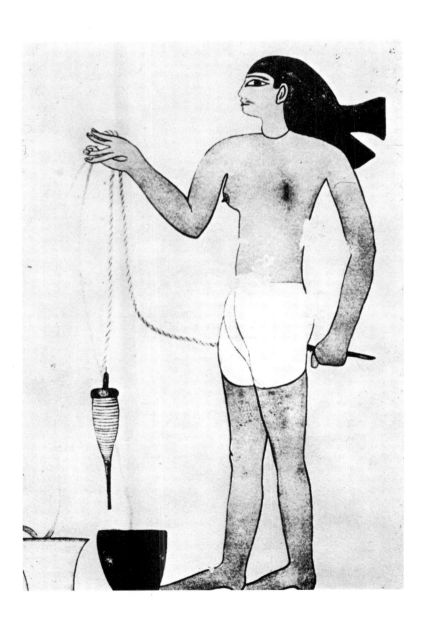

The Middle East led Indo-European civilization in the techniques of measurement as it had done in writing. Standard specimen weights and measures were deposited in temples by 3000 BC. Egyptian measuring cups of about 2000 BC have a respective capacity of one-quarter, one-half, and one and a half *hennu*. A *hon* (the singular of *hennu*) corresponded almost exactly with the capacity of the United States pint, being 477 cc or about 0.84 of the Imperial pint. Length was measured by the cubit which was the notional length of the forearm, and the standard "royal cubit" was flatteringly assessed from the forearm of King Amenhotep I (1546–1526 BC). It measures 523.5 millimeters (20.61 inches) and is divided into seven palms and 28 digits. The palm, at just under 75 millimeters or three inches, contrasts with the western horse dealer's hand, now reckoned at four inches. The Egyptian royal cubit long set the standard for Mediterranean countries. The foot, which originated in Greece at two-thirds of a cubit, stayed stable from country to country, even as far as 12th-century England, between 314 and 317 millimeters (12.36 to 12.48 inches).

Ideographic writing was practised in Mesopotamia, Egypt, China (probably after 2,000 years of slow diffusion from the west), and Central America (where the Maya priesthood of Yucatan almost certainly invented pictography independently around the 3rd century AD). The Sumerians, who had started it in southern Babylonia, simplified their pictorial symbols into formal outlines which could more easily and quickly be marked in clay tablets as cuneiform script.

The Egyptians retained hieroglyphics until after the rise of the Maya civilization, but they reserved the style only for monuments. For documents, for commerce, and for technological communications they developed a cursive hieratic script which could be set down much more quickly, using a brush or reed pen dipped in paint or lampblack ink, and recording on sheets of paper made from the pith of the papyrus reed. Slices of the pith were laid crosswise and backed by the same number of vertical slices before the sheets were pressed, dried in the sun, then gummed into long strips which were assembled compactly in rolls. The earliest preserved written papyri are over 4,000 years old, several centuries after papyrus was introduced. By the 8th century BC the Egyptians further simplified hieratic script for papyrus records into the form called demotic, which was adapted as common utilitarian writing.

Cuneiform, hieroglyph, and demotic maintained their traditional development as ideographic writing. They were never fully phonetic. Eventually, the Sumerians and the Akkadians isolated vowels and represented them by distinct signs, but they never developed signs which identified consonants. (The Akkadians, a Semitic people in the northern province of Babylonia, had conquered Sumer in 2230 BC and adapted cuneiform to their entirely different language.) However, because formalized pictography could be used as an international language, cuneiform characters were used in diplomatic communications between the rulers of Egypt, the Hittite kingdom in Asia Minor, the Canaanite chiefs in Phoenicia, and the Indo-European Mitannite princes and their vassals straddling the subcontinent between the Euphrates and the Mediterranean.

Time was reliably kept from about 1450 BC by the Egyptian shadow clock. The base is engraved at an expanding progression with a scale of six hours. The clock was placed in an east-west direction with the crossbar pointing to the sun – to the east in the morning and to the west after noon. The shortening, then lengthening, shadow divided daylight into the 12 temporal hours which were observed long after the priests of Pharaoh had vanished, within the Christian monastic system and other media.

51

It was in one of the Canaanite capitals, Lakish in Judah, that the earliest alphabetical writing has been discovered, engraved on a dagger.

An alphabet is a series of symbols representing the comparatively small number of individual sounds made in human speech. Analytical writing, which used thousands of arbitrarily chosen outlines to represent ideas, was superseded (except in the Far East) by alphabetical writing because it could be applied simply and universally to the delineation of any language. It was propagated by the energetic Phoenicians, who were active industrially and commercially throughout the then-known world from their strong city states on the eastern Mediterranean north of Mount Carmel. In the first quarter of the second millennium BC the crucial advance was achieved of isolating the consonants in a syllable, creating symbols for them, and evolving a purely alphabetical system. The inscription on the dagger found at Lakish, dated around 1700 BC, is written in a Semitic alphabet in the early Canaanite language.

The successor to this tongue was the North Semitic language, which in its symbols is the most likely source of all other alphabets. The North Semitic language had 22 letters which correspond to the characters in the alphabets of Early Hebrew, Etruscan, Early Latin, and Irish, and are almost paralleled in the alphabet of Early Greek.

In China, writing was not practised until about 1700 BC and, starting as pictograms and ideograms, it has remained unphonetic and unalphabetic ever since. The bone of a specially sacrificed victim from 1600 BC records the questions which the king had instructed the priests to ask the invisible gods, and the answers which the priests had received.

Early Greek took its alphabet, with modifications, from the Phoenicians. There was, however, an "Earlier Greek" alphabet at least 500 years older transliterating a still earlier Greek language. Around 1400 BC the Palace of Minos at Knossos in Crete was destroyed with such sudden violence that, ironically, the equivalent of its waste paper was preserved. Tablets inscribed in a script running between parallel lines – which were actually palace administration chits normally broken up after entry in a master ledger – were abandoned in the ruins. They were not to be seen again for 3,300 years, when they were excavated by Sir Arthur Evans along with a lesser quantity of dissimilar inscriptions, some in a pictographic script, others in a script running in horizontal lines. Evans named the script of the 2,800 tablets comprising his major find as Minoan Linear B. During the next half-century only 134 of these inscriptions were published, as Evans and his legatee Sir John Myres hoarded the works and effectively paralyzed their deciphering.

This cryptographic bottleneck was broken in 1949 to 1952 when Linear B was read by two young enthusiasts, Michael Ventris, a 30-year-old architect, and John Chadwick, a 32-year-old academic philologist. Both had spent the quieter parts of their war years – Ventris as a navigator in the Royal Air Force, Chadwick as an officer in the Royal Navy – studying exotic scripts, including the few available reproductions of Linear B. They declared that Linear B was alphabetically written (in a hitherto unencountered alphabet) in an early form of the Greek language spoken in Mycenae, the mainland city in the Peloponnese which was the center of Aegean civilization from 1500 to 1000 BC. In 1953 detailed work at Pylos in the Peloponnese by the American investigator Carl Blegen confirmed this interpretation. In 1954 Ventris and Chadwick began to design a magnum opus on Mycenean language and civilization. By 1956 Ventris was dead, killed at age 34 in an automobile accident. Professor Sir Alan Wace paid tribute to the laurels he had earned *by deciphering Linear B and discovering the most ancient known form of the Greek language such as was spoken 700 years before Homer.* Ironically, the language of Linear B is not Homeric. It is an inventory of kitchen equipment in the palace which literally reads: *Item, jug, large, with four handles . . .* Writing was still a meagre medium, requiring the courage of countless exploring pioneers before it could be said to reflect that richness of the mind of Man which poets and priests could already convey. But, in fact, the intricacies of science were eventually recorded as well as the subtlety of literature, in spite of the Roscian opposition of the bards who, in Professor Darlington's words, *defended, and for five thousand years have continued to defend, the retreating frontiers of the unwritten word.*

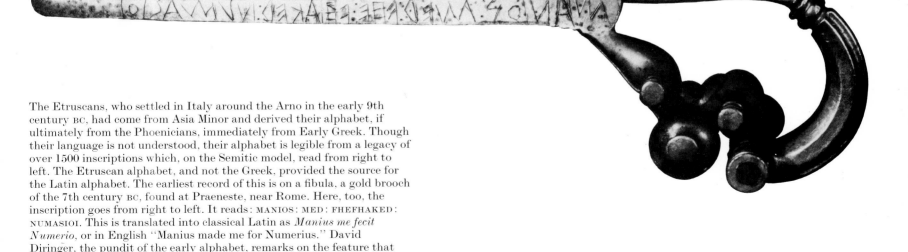

The Etruscans, who settled in Italy around the Arno in the early 9th century BC, had come from Asia Minor and derived their alphabet, if ultimately from the Phoenicians, immediately from Early Greek. Though their language is not understood, their alphabet is legible from a legacy of over 1500 inscriptions which, on the Semitic model, read from right to left. The Etruscan alphabet, and not the Greek, provided the source for the Latin alphabet. The earliest record of this is on a fibula, a gold brooch of the 7th century BC, found at Praeneste, near Rome. Here, too, the inscription goes from right to left. It reads: MANIOS : MED : FHEFHAKED : NUMASIOI. This is translated into classical Latin as *Manius me fecit Numerio*, or in English "Manius made me for Numerius." David Diringer, the pundit of the early alphabet, remarks on the feature that the letters F and h are combined to represent the sound of *f*, which was common in Latin, but was wanting in Greek, and prompted three different devices by which the Etruscans registered this sound.

The first use of copper, found in its natural form, was for shaping to produce trinkets, which have been found at Çatal Hüyük, Turkey and dated around 6400 BC. Later the metal was heated and beaten to change its shape, or even melted and cast into a different shape. Smelting may have been first practiced when copper ores, first used for cosmetics and pigments, were painted onto pottery and baked. From 3800 BC copper ores were mined and smelted and copper tools and weapons were fashioned. Metal tools, not being governed by a grain along which the element split, as with stone, could be shaped in a handier, more functional way. Egyptian copper knives some 4,000 years old have handles which no stone-chipper could reliably have shaped, and the saw retains finely cut teeth which were then unique.

But from about 2000 BC Egyptian smiths, seeking a harder medium, first smelted arsenic with copper and then manufactured the alloy of copper and tin called bronze. Originally this had occurred fortuitously by smelting composite ores. Soon the metallurgists were melting fixed proportions of pure metal, ideally one part of tin to nine of copper. Since Egypt was never self-supporting in the metal supplies it required, excepting gold, tin had to be imported, mainly from Europe. The Egyptians were timid sailors, and the import trade was conducted by bolder navigators, first the Minoans and then the Phoenicians.

Bronze is harder than copper, but its lower melting point made casting more feasible. A bronze two-piece mold from Cambridge, England, was used in the late Bronze Age (around 1000 BC in Britain) to produce socketed axes. A core of clay or sand between the two halves formed the socket-space. Bronze tools were not consistently sharper than stone, but they were more durable and easier to shape. Iron was clearly preferable, but it needed 1,000 years of development after its first exploitation, in about 2500 BC, when it had been worked in its exiguous pure form as a meteorite into rare ornaments and token ceremonial weapons. Smelting iron ore required a far higher temperature – 1,500°C against 1,050°C – than copper, and the wrought iron produced by hammering the slag would not take a keen edge.

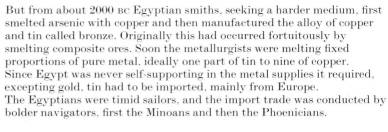

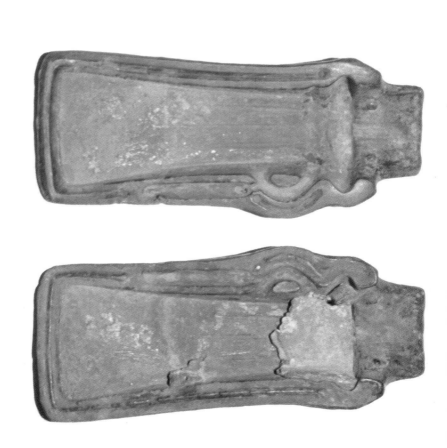

But the tribe of the Chalybes, living southeast of the Black Sea, evolved, around 1400 BC, a process of hammering and reheating the iron in direct contact with charcoal, which effectively produced a steel coating. This process was requisitioned by the Hittites, the overlords of the Chalybes, who then controlled carboned-iron production as a monopoly. The products were keenly sought; as a favor to royalty, the Egyptian king Tutankhamun's father obtained the iron dagger which was found in 1922 in the tomb of Tutankhamun, who had been buried about 1350 BC. The overthrow of the Hittite kingdom in Asia Minor in about 1200 BC dispersed the smiths, who spread the knowledge of iron technology.

At the same time, the Shang dynasty of the kings of China had brought the Chinese – far later than the Middle Eastern peoples – out of the Stone Age into a comparatively high civilization. It was characterized by the introduction of writing and of the light wheeled chariot – the latter certainly taken from more Western models – and by a highly developed bronze technology. The working of bronze was also taken from Western sources but was markedly improved by native refinements. The shafted axe blade surmounted by a human head dates from about 1100 BC. Such a design owes nothing to the stone implements which so closely preceded it. It is entirely functional and, in fact, a similar but broader blade still preserved was earmarked for executions.

Metallurgy enabled more effective tools to be made and more spectacular feats to be done with them. It reinforced the available muscle and mind, the massed brute strength and the occasional ingenuity with which technological problems were solved. Manpower was of sovereign importance at a time when the only other forces which could be reckoned as prime movers were, for limited use, wind, water, gravity and two species of recalcitrant animals. A manpower surplus was both created and demanded by the trend of men towards living in cities, which by 3000 BC had become a notable feature of the river-based civilizations of Mesopotamia, Egypt, India and China. The advanced agriculture on which the cities' prosperity was founded could support more people than were needed to work it: wealth was created. Wealth was increased by activity in new specializations, craftwork and trade. Wealth was stored, counted, appropriated and taxed for the priest-king: an administrative task demanding the labor of many tax officials. Trade and the complexities of an expanding population necessitated the framing of laws and the maintenance of further non-agricultural specialists to administer and enforce the laws, all maintained out of the city's economy.

Trade and taxation demanded accounting, which led to writing, and measurement, which led to science and technological advance. Skilled technical craftsmanship paradoxically created a manpower pool of unskilled but technologically orientated laborers who could, when necessary, be switched by the administrators from navvying in a new irrigation scheme to building a better palace. This reserve of handymen became the first army when the first city – its prosperity still based on the efficient exploitation of an agricultural economy – decided to augment its prosperity by acquiring fresh agricultural heartland through war. War was the creation of cities, which were the first organizations of humans with the capacity to summon, administer, equip and victual large bodies of men in the field. *War, in its full sense, is indeed a product of civilization*, said Professor J.D.Bernal. *The fighting which recurred between tribes in the hunting or even in the pastoral stage was more in the nature of football matches than of sustained campaigns.* No early civilization pursued war more assiduously than the Assyrians, *notorious for their insensate destructiveness and ingenious sadism*, as Lewis Mumford described them, comparing them with the later Mongols and Aztecs.

The Assyrian Empire of 744–609 BC, which stretched at its zenith from the Nile to the Persian Gulf, based its power on hardened steel weapons, advanced military tactics including the use of cavalry, and technological innovations exploiting iron: the use of armored wagons, heavier chariots, and siege weapons which included grappling irons and wheeled siege towers. The two-decked battering ram depicted in low relief on a carving found at Nimrud, near modern Mosul, had an iron head to the ram which was manipulated by sappers inside the hide-covered frame to detach the stones from walls while archers on the top deck harassed the defenders. The machine was manhandled into position by soldiers. War gave impetus to the cultivation and application of science in an age when technical advance in other spheres had lapsed into decline. The profession of *engineer* was instituted, with mainly military connotations. Yet the Assyrians used their new engineering mastery on civil projects, dams and aqueducts and the colossal buildings erected in Nineveh. Their addiction to the use of stone involved them in the problems of mass-haulage and leverage which are suggested in the sculpture from Nineveh of a workforce erecting an obelisk.

Classical Greece preferred to be noted for its attachment to theory rather than practice and to science rather than artifice, yet it depended heavily on technology and treated its artisans well. By 510 BC legal reforms had given craftsmen in Athens the status of citizens, though they were often of immigrant or slave origin. Attica needed an immense reserve of productive mechanics to manufacture the exports of metal and metalwork, cloth and pottery as well as oil and wine, with which it paid for its imports of food and luxuries. Ironworking, a skill inherited from refugees from the break-up of the Hittite empire, is graphically illustrated in the smithy portrayed on a Greek amphora from about 500 BC found at Orvieto. The master smith holds the workpiece as his assistant swings the hammer before seated onlookers. The range of tools is impressive. The artistic force of the design conveys the godlike awe which early smiths inspired, giving them a warm corner in mythology.

China had had dynasties of emperors from 2200 BC, 900 years after the first Egyptian line of Pharaohs. China's second dynasty, the Shang, represented the migration from western and central Asia of a technologically competent, caste-conscious people who were efficient farmers, skilled metalworkers (in bronze), and intrepid fighters, exploiting the new military arm of the horse-drawn chariot. The mobility given by the horse facilitated the swift Shang domination of China but it presented the conquerors with the threat of a new wave of hostile horsemen coming east from central Asia. In the last two decades of the 3rd century BC, between 221 and 202, the initially powerful Ch'in dynasty unified China by ruthless military oppression and built the 1,500-mile Great Wall to keep out the Hun horsemen invading from the west and north. Basically the wall was up to 30 feet high, tapering from a width of 25 to 12 feet, with 40-foot-high towers spaced at 200-yard intervals. Horses could not surmount it. The Hun invaders had to dismount, and were easy targets for the new Chinese bronze-lock crossbow. This ultimately effective defense turned the warlike migrations of the Huns to the west.

The Han dynasty of China, which on the collapse of the Ch'ins restored order for 350 years beginning in 202 BC, presided over the technological and intellectual flowering of the country, even establishing land and sea communication with India and Rome. The seismograph of Chang Heng, dated 132 AD, which is reconstructed and dissected here, is of bronze. An earthquake shock swung an internal pendulum which dislodged one of eight horizontal arms. This caused the mouth of one of the hanging dragons to open and release a ball which fell into the mouth of the toad below, indicating the direction of the earthquake shock. The chamber is eight (Chinese) feet in diameter. The contemporary model of a Han dynasty grain mill has a pedal-operated tilt hammer for beating the husks off grain and a rotary winnowing mill worked by a crank handle.

Through slave labor and an enormous army, the Romans efficiently tackled large-scale engineering as a nationalized enterprise. Their resources were almost infinite and their technological skill, recruited imperially and exploited to the full, made the best use of their equipment. During the five centuries after 300 BC they ran a web of roads for thousands of miles through Europe and the Empire, patiently excavating the trench in the superbly surveyed line of the road, filling the trench, laying the roadbed, which they surfaced, paved, and lined with curbstones. On slippery mountain passes they dug special ruts in the rock to hold cartwheels. (The Greeks had relied on engineered road ruts, like inverted railroads, curiously with a common gauge of 138 centimeters, comparable to the present standard rail gauge for Europe and the United States.) Roman aqueducts ran mainly underground, though long raised stretches keeping up the gravitational level of water were impressive. In building, the Romans placed surprising reliance on concrete, even for architectural domes, using a cement based on volcanic earth with an aggregate of rock and pumice.

In the 1st century BC Roman engineers improved the water mill, the machine that from classical antiquity had turned natural forces to human needs. The Greeks had used the water mill only as a horizontal power wheel. Now the Romans utilized Greek gearing to make a vertical wheel rotate horizontal millstones to grind grain – efficiently, with five revolutions of the millstone to one of the wheel. The water mill was, however, surprisingly little used, except in some areas of Roman France. In a simple stepladder elevation on a watercourse near Arles 16 water-wheels progressively using the same flow drove 32 mills to produce 28 tons of meal in 24 hours. The water mill was described by Vitruvius and is therefore often ascribed to him. Vitruvius, a Roman military engineer of the 1st century BC who served under Julius Caesar, wrote an invaluable handbook of mechanical engineering, and did indeed make notable innovations. He built and described a crane for lifting heavy objects, using a triple pulley. An example of the Vitruvian crane is shown in a relief in the Lateran museum dated 100 AD. Five men on a treadmill hoist material to build a monument, applying their force through compound pulleys rigged to a single mast supported by stays.

As a means towards *the multiplication of forces* the pulley had been known in the 3rd century BC to Archimedes of Alexandria and Syracuse, the reputed inventor of the "Archimedean screw" for lifting water from one level to another. In fact, Egyptian peasants had raised canal water by this means to irrigate their fields long before Archimedes, and still do today. The mechanical hoist is an exploitation of a prime problem in mechanics: how to convert a small force acting over a long distance into a large force acting over a short distance. The 1st century AD theorist of mechanics was Hero of Alexandria, who standardized the list of mechanical elements which allow this transfer of energy: the windlass, the lever, the pulley, the wedge, the screw. All five of these "means towards the multiplication of forces" were present in the simple and satisfying siege weapon, the *ballista*, the heavy artillery catapult developed by the Greeks and enthusiastically put out for multiple production by the Romans. Hero inherited the traditions of, and paid intellectual homage to, Ctesibius of Alexandria, who almost two centuries earlier had ingeniously improved the water clock and was responsible for a number of mechanical toys.

Hero's approach to mechanics was similarly to treat it largely as a toy for the display of near-magical wonders. He built a mechanically working puppet theater. He made a pneumatic/hydraulic door opener for priests to use in the temple to impress the devout – the doors opened for as long as a sacrificial fire was lit. He detailed the instructions for making what was called "Hero's fire-engine" using the principle of the force pump introduced two centuries earlier by Ctesibius. A two-cylinder force pump moved water into a chamber where it came under increasing pressure from the air being compressed by its entry. The outlet to the chamber was a flexible tube joining it below the water line. Utilization of the cushion of air kept up the water pressure so that the liquid was ejected in a steady stream instead of with separate spurts. After the Dark Ages Hero's instructions were mislaid, and when the fire pump was re-invented in 1615 it merely shot intermittent squirts until the air chamber was incorporated 40 years later.

Hero's most significant innovation was the construction of a hollow sphere with two tube outlets in its upper half. Water was boiled in the chamber and the escaping steam, reacting in the tubes as two jets, whirled the sphere around like a top. This was a true steam engine, converting steam power to motion, but the principle was virtually ignored until Newcomen.

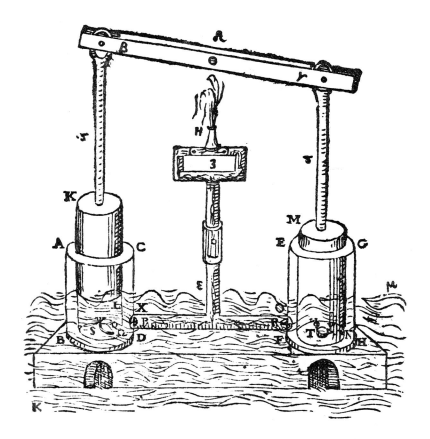

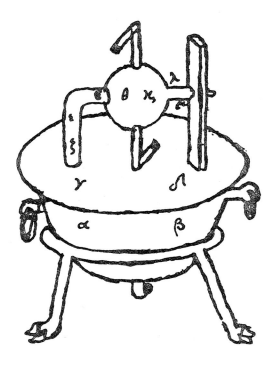

On March 27, 196 BC – the 4th Xandikos and the 18th Meshir of the dynastic year 9 – the priests of Memphis drew up a testimonial in praise of their 26-year-old Pharaoh, Ptolemy V Epiphanes (the great-great-grandfather of Cleopatra). The priests decreed that the memorial should be engraved in the three public languages then used in Egypt: hieroglyphics, Greek, and the cursive script of Egyptian demotic, the informal everyday language of the country. Twenty centuries later, on August 2, 1799, a French soldier dug up a stone bearing a triple inscription near Rosetta, a vantage point on one of the effluences of the Nile. Napoleon had become the temporary Pharaoh of Egypt, not only for military and commercial ends, but to gain intellectual honor for France as the patron of science and culture. His 350-strong convoy to Egypt were crammed not only with military personnel, but with engineers, architects, and professors of every specialty. Their passenger list, according to the English Admiral Collingwood, ranged *from astronomers to washerwomen*.

A description of the Rosetta Stone was in print in the *Courrier de l'Egypte* within four weeks of its discovery. Many plaster casts were taken before the Rosetta Stone was surrendered to the British as a spoil of war in 1801. Of the three inscriptions on the stone, only the proclamation in Greek was understandable. Clearly it was the clue to the decipherment and understanding of hieroglyphics and demotic. Between 1814 and 1822 Jean François Champollion virtually completed the decipherment, but his exposition of the languages was not accepted. In 1866 Richard Lepsius, who had already perfected the work of the long-dead Champollion, discovered at San in Egypt a further trilingual inscription, the Canopus Decree, which he could read on sight. As with the script of Linear B and the secret of the AntiKythera computer, many of Man's early achievements have remained mysteries which have been solved only by the skills of men living thousands of years later.

The first practical uses of the calendar were agricultural: by forecasting the seasons accurately farmers could schedule the planting of seeds in the spring, and could appease the gods by appropriate celebrations. For many centuries the calendar was regulated by the lunar cycle. The mathematical rule, known long before by the Babylonians, was enunciated in 432 BC by the Greek astronomer Meton. It was based on the lunar month, the period of some $29\frac{1}{2}$ days between one new moon and the next. Meton showed that 235 lunar months make exactly 19 years. The Greeks constructed a lunar calendar (which has been retained by the Jews and is utilized by the Christians to ascertain Easter) in which cycles of 12 years of 12 lunar months and seven years of 13 lunar months completed 235 lunar months. Every 19 years the seasons would coincide exactly with the calendar.

The Egyptian priests abandoned the Moon for the Sun, said the Earth made one revolution round it in 365 days (their astronomers knew differently), and adopted a solar calendar having 12 months of 30 days with five extra days in the year, making the calendar one day slow after every four years.

The ancient Romans, who mixed myth with history, said that Romulus instituted a year containing 10 lunar months. Clearly, even after one year, this produced a useless calendar.

Numa added the months of January and February to come before March, which pushed September, October, November, and December from their nominal positions as the seventh, eighth, ninth, and tenth months. The 12 lunar months were alternately of 29 and 30 days, with the addition of one for luck because an odd number was thought to be a better augury, making 355 days in the year. Since there were still some $10\frac{1}{4}$ missing, Numa put into every second year, between February 23 and 24, an intercalary month of 22 days alternating with 23. This was an overcompensation, making the average year one day too long. When it was realized, 24 days were knocked out of the calendar every 24th year. It was a system that led not only to confusion but to a novel form of corruption. Since the political pontiffs controlled the insertion of the intercalary months, with power to add or subtract further days, they would "adjust" the year either to prolong the power period of an administration they wished to continue or to shorten the period of office of political leaders they wished to remove. In addition, Roman legislators were poor mathematicians. The consequence was that by the time of Julius Caesar the calendar was in such a hopeless mess that if any farmer decided to do his spring sowing with reference to the date rather than the weather, he would be planting his seeds in midwinter.

In 46 BC Julius Caesar revolutionized the calendar after consultation with Sosigenes, an astronomer and mathematician from Alexandria. Sosigenes confirmed that the solar year had a duration of $365\frac{1}{4}$ days. (He was $+0.007801$ days in error, which led to the Gregorian correction 1,628 years later.) Caesar abolished the year based on the lunar month, with the intercalary months which it necessitated. The year was to have 365 days with one extra day in every fourth year. The odd months of the year were to have 31 days, the even months 30. February was to be docked one day, giving it 29, for three years out of every four, to make the total correct. Since the calendar was already 90 days out of true, 46 BC was, as *the last year of confusion*, to have 90 extra days to get right the midwinter festival of the Saturnalia and the consequent vernal equinox of 45 BC. It was a logical system of reasonable accuracy. It began to crumble within two years. Caesar was assassinated in 44 BC, and a contrite Senate voted to call the seventh month after the martyr: the English July. Thirty-six years later the Emperor Augustus demanded a similar honor and the eighth month was called August. Egotistically, he insisted that his month should have as many days as July. August was given 31 days by expropriating a day from the much-put-upon February.

But since it was considered unlucky to give 31 days to the consecutive July, August, and September, September was docked of a day which was given to October, and November 31 was allocated to December. Things might have gone well, if difficult to memorize without a nursery rhyme, had not the Roman pontiffs failed again to grasp the simple mathematical rules of Sosigenes and Caesar. With faltering calculation, they made every *third* year a leap year instead of every fourth. By the time Augustus claimed August, the calendar was already three days fast, which the emperor corrected by abolishing leap years for the next 12 years. This took the calendar into the Christian era, which was not in fact designated as such until 533 AD. Owing to imperfect chronology, this year was actually only 529 years after the birth of Jesus Christ.

In the year 80 BC a ship bound for Rome from the island of Rhodes struck rocks and foundered off the Aegean island of AntiKythera. The wreck was not to be discovered for 1,980 years, when divers brought parts of its cargo and equipment to the surface. Among this treasure trove was what Professor Derek J. de Solla Price has described as *the remains of the most complex scientific object that has been preserved from antiquity.* It was a box, roughly square in two dimensions but standing on a narrower base. On the square face of the box there had been dials. Around them there had been engraved, and there still partly survived, long technical descriptions and instructions in Greek. Inside the box was a very sophisticated assembly of at least 20 gear wheels mounted eccentrically on a turntable. The recovery of gears of this complexity, reputedly from 80 BC, was long treated with skepticism. Previously the most elaborate gears developed by the Greeks had been believed to be those of the hodometer.

After decades of conjecture, research, and meticulously delicate cleaning by the Greek National Archaeological Museum in Athens, the inscriptions could be read or surmised by the epigrapher George Stamivas. Professor Price identified the mechanism as *an arithmetical counterpart of . . . the geometric models of the solar system . . . which evolved into the orrery and the planetarium. The mechanism is like a great astronomical clock without an escapement, or like a modern analogue computer which uses mechanical parts to save tedious calculation.* The main section is a bronze plate on which were mounted gear wheels primarily driven by a crown-gear wheel. It turned on an axle which, Price speculates, may have been kept in motion by a mechanism similar to a water clock. The crown wheel moved the main four-spoke driving wheel, still obvious with its 240 teeth. This was connected with two trains of gears which turned gear trains leading through a differential system to a set of shafts turning the various dial-pointers on the face of the machine, marked with the signs of the Zodiac and the months of the year. They showed the annual motion of the Sun in the Zodiac; the main risings and settings of bright stars and constellations throughout the year; the main lunar phenomena – the phases of the Moon and the times of rising and setting – and useful astronomical information on the planets then known to the Greeks – Mercury, Venus, Mars, Jupiter, and Saturn. Professor Price decided that the instrument had been made probably in 82 BC. He observed that it was adjustable for leap years and had in fact been reset in 80 BC, the year the ship is thought to have sunk off AntiKythera.

The instrument is in fact astonishingly similar to astronomical clocks which were in great vogue during the Renaissance in Europe. *Behind the astronomical clocks of the 14th century,* declared Professor Price, *there stretches an unbroken sequence of mechanical models of astronomical theory. At the head is the AntiKythera mechanism. Following it are instruments and clocklike computers known from Islam, China, India, the European Middle Ages. The importance of the line is very great because it was the tradition of clockmaking that preserved most of man's skill in fine mechanics. During the Renaissance the scientific instrument-makers evolved from the clockmakers. Thus the AntiKythera mechanism is, in a way, the venerable progenitor of all our present plethora of scientific hardware.*

THE DECLINE AND FALL

Overleaf: rubbing from the tomb of Wu Liang, 147 AD, showing improved horse harness

In 381 AD Ambrose, bishop of Milan, described the towns of Central Italy as *remains of half-ruined cities*. In the summer of 410 Rome fell to a Visigoth army; and St. Jerome, who was living in Bethlehem, and not yet a saint, wrote: *The light of the whole world is extinguished. If Rome can perish, what can be safe?*

The Roman world was a world of towns, schools and universities, of writers and builders, of workshops, gymnasia, propylaea, and temples. In the barbarians' world – as one historian put it – *mind was in its infancy and its infancy was long*. The fall of Rome ushered in an age of barbarism and obscurantism in the West destined to last for more than half a millennium. We can discuss interminably whether, as another historian once put it, *Rome was killed or died a natural death*. Whatever the causes of the decline and fall its consequences were ominous and they remind us that civilization and its concomitant technology are precarious and perishable: they cannot be taken for granted. From the 5th to the 12th centuries Western Europe, which we identify today as one of the main centers of civilization, was primitive, not merely by the standards of modern industrialized societies but in relation to the China of the T'ang and Sung Dynasties, the Byzantine Empire of the Macedonian Dynasty, the Arab Empire of the Ommayads and the Abbasids.

In 807 AD Haroun al Rashid sent an embassy to Charlemagne, along with several presents, among which was a water clock. Nothing of the kind had ever been seen at the Frankish court, and its description by a Carolingian chronicler reflects the astonishment and admiration the contraption aroused. Nobody in the West could have fabricated such a wonder. In 949 AD when Liutprand of Cremona visited Constantinople and was admitted to the presence of the Greek Emperor, he marveled at the mechanical contrivances which made the throne rise up, and the bronze lions beat the ground with their tails and *give dreadful roars with open mouths and quivering tongues*. As late as the beginning of the 13th century, nothing in Western literature could compare with al-Jazari's technological encyclopedia. When Marco Polo reached China he observed with amazement that pieces of paper there served as means of payment.

For about half a millennium after the fall of Rome, the Orient produced the world's most refined artifacts and the most advanced technology. The West mostly did no more than adopt and adapt ideas.

The availability of paper, and the consequent development of printing, has been the sluice permitting the sure, smooth flow of communication, education, and the dissemination of thought. For some 800 years until the 7th century AD the technology of papermaking was confined to the Chinese Empire and adjacent lands. It was acquired by the Arabs in the 8th century and only tardily passed on to Europeans in the 13th century. During all that time, the materials and method hardly changed after an early refinement. *The manufacture of paper was sent forth from the Chinese dominions as a fully developed art*, said Thomas Carter, the pioneer investigator of the subject. *Paper of rags, paper of hemp, paper of various plant fibers . . . paper of various colors, writing paper, wrapping paper, even paper napkins and toilet paper – all were in general use in China during the early centuries of our era.*

The first books in China were written, with pen and ink, on vertical strips of wood or bamboo, often corded together at one end around the wedge of a "bottleneck." By the 3rd century BC they were being brushed onto rolls of silk. Chinese records, which are peculiarly reliable as history, state that, *silk being too expensive and bamboo too heavy*, a method of papermaking evolved *using tree bark, hemp, rags and* [used] *fish-nets*. The date of this innovation is put prior to 100 BC. By 105 AD the method had been perfected. The earliest existing Chinese paper was found in Turkestan, dated 150 AD. Over the following years the paper was improved for greater strength and better absorption of the ink, typically by impregnation with a size of starch. In the 8th century, when the Chinese had already developed block printing, the secret of papermaking was taught by Chinese prisoners of war to their Arab captors in Samarkand. Subsequently, Damascus became the major center for the export of paper to the West. When the Arabs conquered Spain, they established papermaking in Europe, notably at Xativa in the province of Valencia, where by 1151 energy seems to have been provided by water mills. In 1276 the Italians were exploiting water power for papermaking at Fabriano, and the craft spread very slowly through Europe. It reached England around 1494 and Russia – via Western, not Oriental sources – in about 1565, more than 16 centuries after the inception of the technology in China.

In all lands west of China, which had exclusive use of paper until 751 AD, extended writing was done on leaves, linen, animal skins, or papyrus. The Egyptian *Book of the Dead* was written on leather around 3000 BC. Very shortly afterward papyrus began to be used, eventually glued into rolls from 30 to 150 feet long. Parchment – sheets of fine animal skin which is not tanned, but dried and shaved – was at first a specialty of Pergamum, a center of Hellenistic culture near modern Izmir. The best quality, made from the skins of new-born calves, was called vellum. Ink was prepared from carbon, or by treating a salt of iron with gallic acid from oak galls: it was applied in a sandy color and became black through oxidation from the air. In the 2nd century AD parchment began to be cut into rectangular sheets, sometimes folded into folio, quarto, or octavo, which were stitched or glued together at one end and bound into a volume, the codex. The sheets of a codex could be used alternately on recto and verso. Very long texts such as the Bible could be completed. Reference to a passage was far easier than with the cumbersome roll. And vellum ousted the more brittle papyrus for "eternal" works. When the Emperor Constantine accepted Christianity as the official religion of the Roman Empire around 320 AD, he ordered numerous copies of the Bible to be written (in Greek) on codices of vellum. The *Codex Sinaiticus* dates from shortly after this decree. The 390 remaining leaves measure 15 by 13½ inches and are mainly written, in iron ink, with four 48-line columns to a page.

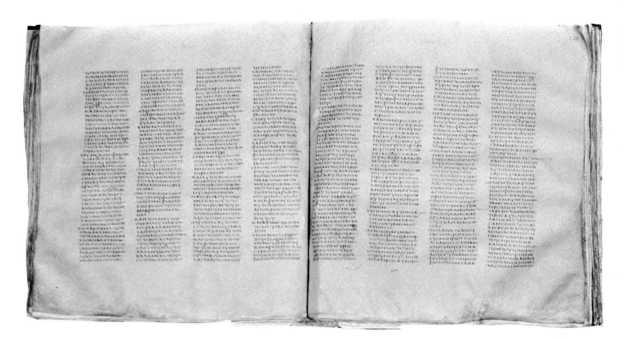

During the four centuries of the Han dynasty in China, straddling the beginning of the Christian era, engraved seals were used increasingly as emblems of personal prestige and official registration. Around 500 AD these seals began to be cut in relief so that, when inked with vermilion, the design showed as a red impression on a white background. By the end of the 7th century a particular adaptation of seal stamping was the consecutive impression of figures of Buddha, printed as "holy pictures" at the top and bottom of every column of a manuscript. The stamps which have been recovered, made from metal as well as wood, are typically some six centimeters high and were individually inked and pressed by hand. They mark the transition in China from seal impression to wood-block printing.

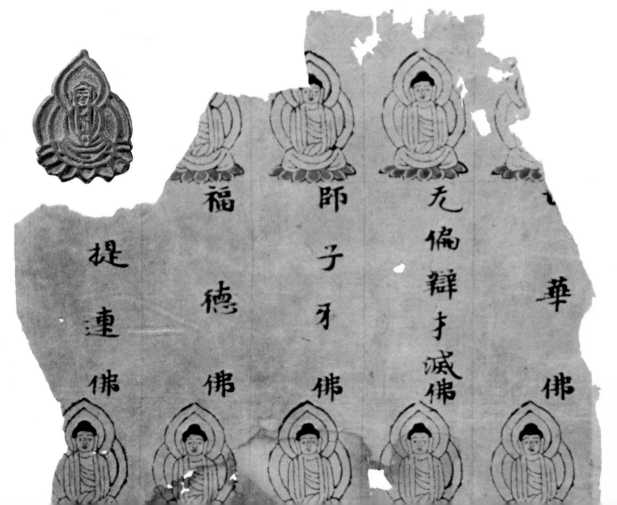

The oldest known printed book in the world was found in peculiar circumstances. A wandering Taoist friar had begged enough money to restore the frescoes in one of the shrines in the Caves of the Thousand Buddhas, which were cut into a cliff in the desert of Chinese Turkestan and were for 1,600 years a place of Buddhist pilgrimage. He cut too deep into the plaster and exposed brick instead of stone. Sacrificing the fresco, he dug out the brick and came through to a rock chamber some nine feet square. Inside this cell were about 15,000 manuscript rolls, written on paper, dating between 406 and 996 AD, and perfectly preserved though the year was 1900. They were mainly in the Chinese language, but included work in five other ancient Asian languages and a book written in Hebrew containing excerpts from the Old Testament. Among the manuscripts was a printed book, a roll containing seven sheets pasted together. Six of the sheets were printed text, each sheet (and therefore the blocks which had been cut to print them) over 24 inches long by 12 inches high. The seventh sheet was an almost square woodcut showing Buddha expounding his teachings to his disciple Subhuti. The work is the *Diamond Sūtra*, an important book in the canon of the Buddhist scriptures. The date of printing is precisely given in a colophon preceding the illustration, and repeated on a filing tag pasted on the outside of the roll. The imprint reads: *Reverently made for universal free distribution by Wang Chieh on behalf of his two parents in the 15th of the 4th moon of the 9th year of Hsien-t'ung* [May 11, 868],

Significantly, the first known printed book had been distributed freely as popular culture, or as ideological propaganda. What is of deeper technological impact is the amazing quality of the product, printed on paper 586 years before the Gutenberg Bible, block-printed with a sophistication which implies many decades of previous refinement of the craft, the woodcut was designed with a graphic understanding that could never be commissioned by, say, Alfred the Great, who was translating ecclesiastical philosophy for written transmission at roughly the same time. Though the graphical technology of the *Diamond Sūtra* is of mature development, there is evidence that it is among the earliest literary works printed. Some of the publications found there can be reliably dated from the next century. Yet among all the 15,000 rolls only four others are printed roll books, and there is one printed folded book dated 949. But there were some scores of single-sheet block prints, cheap religious pictures about half the size of the *Diamond Sūtra* illustration. Evidently block printing of single illustrations was the inception of the new technology. There is indeed an archive dated 835 prohibiting the printing and sale of wood-block calendars as an unlicensed form of private enterprise in China. Yet there is a record, almost contemporary with the publication of the *Diamond Sūtra*, indicating that an official named Ho-kan Chi *composed the Biography of Liu Hung and had several thousand copies printed, which he sent to those who . . . practised alchemy.*

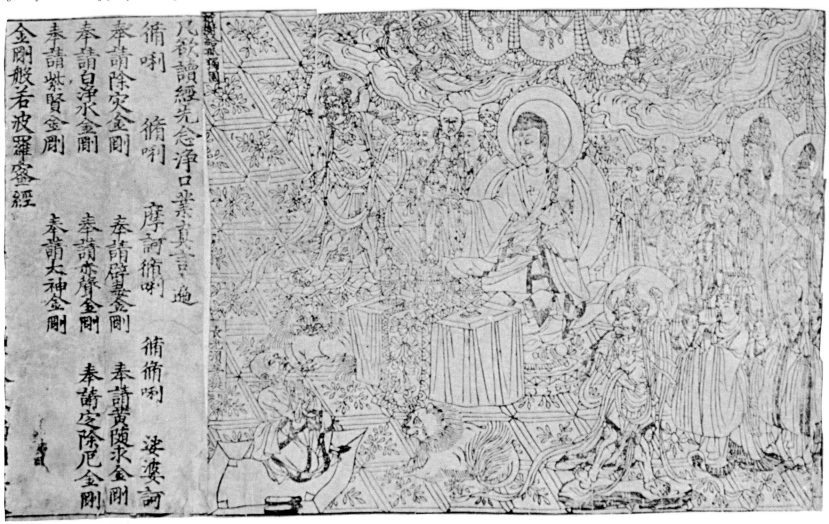

The horse originated in the steppes of central Asia and was used as a mount or in draft (which for centuries implied pulling the war chariot, since adaptation to agriculture or trading transport was slow). The mounted warrior initially rode, using only the pressure of his knees for guidance and control, leaping onto and off his animal as circus riders do today. Reins attached to a bit came later.

From about the 4th century BC the horsecloth, which had been adopted earlier, began to be padded fore and aft of the rider's torso, and this developed into the made-up saddle which was taken over from the 1st century AD by both the Chinese and the Romans. But a horse-soldier, though he had fore-and-aft stability, still had no lateral support and could easily be jerked off sideways. He was a man on a horse,

using his beast for mobility but not impetus. In a fight he was a bowman or spearman with advantages of height and speed. When he used his spear he employed his own strength to direct, deliver, and extract his weapon. (The last operation was a particular problem until a baffle below the spearhead was fashioned to prevent too deep penetration.)

What eventually gave the rider lateral control was the introduction of stirrups. Probably originating in China, they were used in Korea, Japan, India, and Iran before they were taken up in the West. It was the Franks under Charles Martel, the grandfather of Charlemagne, who exploited the stirrup in Europe from about 732. The stirrup, clearly seen in expert use by the cavalry of Charlemagne in the late 8th century, enabled the resources of man and horse to be combined by using the full impetus of the animal and creating the new fighting arm of true cavalry. In the words of Professor Lynn White, the stirrup *effectively welded horse and rider into a single fighting unit capable of violence without precedent*. By combining human energy with animal power, it *immensely increased the warrior's capacity to damage his enemy . . . It made possible mounted shock combat, a revolutionary new way of doing battle.* Additionally, because it led to the development of heavier armor, it provided the incentive for breeding a heavier type of horse, which in turn made the animal more suitable for agricultural labor.

Fire had been used as a weapon of war since the first flaming torch was thrown into the first hut. "Liquid fire" was a weapon peculiar to western Asia, the only area in the then-known world where the combustible natural petroleum product known as naphtha seeped from the ground. By about 500 BC compounds of naphtha and bitumen were being assembled into fire pots and discharged as missiles from the torsion catapults called *ballistae*. At the siege of Rhodes in 305 BC 800 of these artillery engines were used during one night-long assault. The ballistic missiles did not explode or ignite spontaneously, but were lit immediately before being projected. In the 7th century AD an improved liquid combustible based on naphtha was developed within the Byzantine Empire, and was decisive in securing the survival of Constantinople against the Muslims between 673 and 678, and against later assaults from the West. Known as *Greek fire*, it remained a secret weapon for centuries before its constituents were disclosed.

A volatile variant of Greek fire was prepared for naval warfare, and was illustrated in a Byzantine manuscript of the 10th century. It was, in fact, made from naphtha, powdered quicklime, sulphur, and possibly saltpeter (potassium nitrate), and it burned on the surface of sea water. It was shot as a burning jet from a bronze siphon force pump fitted to the prow of a ship. The Muslim armies, using their supply base of natural petroleum seepages in Iraq, developed naphtha fire pots as a weapon of war, and the West adopted the use of the missiles, which were only supplanted by mortar bombs in the 16th century. Naphtha-based incendiaries were to be revived in 1944 with the development of napalm, the jellied petrol made from naphthalene and palm oil.

The practical water clock was introduced to Alexandria in the 2nd century BC by Ctesibius, who adapted the ancient clepsydra from a simple cylinder from which water poured to a drum into which water steadily dripped, raising a float which marked the time in clear gradations. Yet simple water clocks were also in use in China by 100 BC, being constantly watched by slaves who beat out the hour on a drum when the cistern emptied. By the 6th century AD the Chinese were building elaborate water clocks, probably based on Hellenistic models, but they had adapted water wheels to harness power to drive the considerable mechanism then involved. By about 724 an engineer named Liang Ling-tsan, constructing a huge clock which included a moving celestial sphere with a separate Sun and Moon, introduced the escapement. This mechanism permits finer regulation and slows down the speed of the moving parts – in this case governing the speed of rotation of the water wheel through tripping lugs. In 1090 Su Sung built the most monumental Chinese water clock. Its escapement slowed the driving wheel to one revolution in nine hours. A crown gear from the driving shaft worked pointing hands, bells, gongs, and drums which marked different hours. In the 35-foot-high tower the moving celestial sphere carried pearls to denote the stars. The clock worked for some two centuries.

The astrolabe, an instrument for taking the altitude of heavenly bodies, was first developed by astrologers and only later taken up by navigators and "pure" astronomers. In his *Treatise on the Astrolabe* Chaucer told *litel Lowis my sone* of descriptions of the astrolabe written in Latin, Greek, Hebrew, and Arabic. The earliest surviving instrument, one of many fine examples of Islamic art in this field, dates from the 10th century and was made in Isfahan, which is still a center for astrolabe craftsmanship. Arabic astronomers went on to develop mechanical astrolabes, using gear mechanisms to place the stars and planets correctly. These instruments, spreading into Europe, were the forerunners of the mechanical clock.

An intricate hydraulic system was the prime mover of the Palace Water Clock described in 1205 by al-Jazari of Diyār Bakr, on the northwest border of the ʿAbbāsid Islamic empire based in Baghdād, near modern Erzurum in Turkey. Al-Jazari wrote a famous encyclopedia of automata, *The Book of Knowledge of Ingenious Mechanical Devices*, which fully described 50 complicated automatic wonders and gave specifications for their construction. Many of them, like the automata of Ctesibius of Alexandria, were toys for princes or extravagantly bizarre gadgets, like a cocktail dispenser or a mechanical hand washer. However, the Palace Water Clock was a practical if excessively ornate device built like a palace façade, in which the passing of time was noted by windows opening, musician dolls playing, birds singing and dropping balls from their beaks into vases. Above the roof was a copper disk marked with the signs of the zodiac: Cancer was illustrated at the top and was followed, counterclockwise, by Leo, Virgo, Libra, Scorpio, Sagittarius, Capricornus, Aquarius, Pisces, Aries, Taurus, and Gemini. Within this disk a separate disk revolved with the passage of the Moon, and the inmost disk depicted the sphere of the Sun. The illustration, carefully drawn in the extant 1315 edition from al-Jazari's original, is one of 173 beautifully executed engineering diagrams accompanying the commentary.

Ingenious Mechanical Devices is but a showpiece of the intense Islamic preoccupation with science, mathematics, medicine, and technology. The Arab empire had access to sources – for example, some of the Greek works of Hero of Alexandria – which were eventually lost. Many Greek treatises became available to Europeans of the Renaissance only through Arabic translations. The cultural pioneer was Frederick II of Sicily, Holy Roman Emperor, who founded at Naples in 1224 the first chartered university in Europe, and deposited there a mass of Arabic documents including translations from Aristotle. Thomas Aquinas studied at Naples 20 years after its foundation.

In many branches of science, but particularly in medicine, Islam took over a heritage of knowledge from India, Persia, and later from China, besides the store of tradition which descended from Alexandrine Egypt by way of the assiduous translators from Syria. Euclid, Galen, Ptolemy, and Aristotle were all translated, by order of the Caliph, first into Syriac and thence into Arabic. Seven of Galen's books on anatomy, extinguished in their Greek original, were preserved in Arabic. Hindu arithmetical lore, with its use of the numerals (which we now call "Arabic numbers"), was introduced to Islam between 867 and 874. Only slowly accepted there, the numerals took even longer to be adopted in Europe. The oldest dated European manuscript using Arabic numerals (though still lacking the zero) was written in Spain in 976. It is shattering to reflect that a Western mathematician at the time of William the Conqueror, asked to square 99, would have had to project LXXXXVIIII times LXXXXVIIII before he even began to compute the product.

THE URBAN REVOLUTION

As late as 1000 AD Western Europe was a most primitive, depressed, and stagnant area, underdeveloped not only by modern standards but also by those of contemporary societies such as the Byzantines, Arabs, and Chinese. Violence, superstition, and ignorance prevailed. The people were few, short, and illiterate; the great majority lived primitively and poorly in the countryside. The cities lay in ruins. Trade relations over long and short distances were scarce, irregular, and unpredictable. Those few exchanges that took place were largely carried on by way of barter. The structure of the society was characterized by a tripartite division between those who fought, those who prayed, and those who worked. Those who fought did it mostly in order to rob. Those who prayed did it superstitiously. Those who worked were regarded as serfs and occupied the lowest steps of the social ladder. The state of the arts was in line with the primitiveness of the social structure.

By the year 1200 AD the situation had totally changed. Population, production, trade, the use of money, the state of the arts, economic opportunities were all expanding and Western Europe was rapidly catching up with the rest of the civilized world. In fact, it was laying the foundation for its future predominance.

The cities were the engine of change. They grew in number as well as in size and wealth after the middle of the 10th century, and as they grew they gave European development its most distinctive characteristics.

Overleaf: Medieval city gate of York, England

Cities existed elsewhere in the world, as they had existed in classical antiquity. But the city of medieval Western Europe was a totally new and unique phenomenon. The traditional city was an organ of a larger organism, dominated by the cultural values and political influence of the landed gentry. The cities of medieval Western Europe grew as organisms in their own right, fiercely autonomous both politically and administratively. A person who passed through the gates of a medieval city became subject to different laws, as one does today when one crosses the border from one country to another. The profound significance of the urban revolution of the 11th and 12th centuries stemmed from the fact that the medieval cities nurtured within themselves an autonomous culture totally foreign and indeed antagonistic to the surrounding feudal environment.

The representatives of the landed feudal aristocracy had no sympathy for those new societies made up of burghers. *Communio: novum ac pessimum nomen* (Commune: new and despicable name) thundered a conservative of the caliber of Guibert of Nogent. But the feelings were mutual. In the cities the cultural atmosphere was decidedly antifeudal; and in Northern Italy, where urban growth was particularly vigorous, the cities often engaged in military expeditions aimed at the destruction of the feudal power in the countryside.

In Western Europe the town was the 'frontier', a new and dynamic world where people felt they could find ample opportunities for economic and social advancement. A massive migration then took place from the countryside to the cities. Towns became filled with people who left behind a rural and feudal world they fundamentally despised. City walls were functional, but they also acquired a symbolic significance: they marked the boundary between two antagonistic cultures. The growth of the cities meant the growth of literacy, learning, skills, production, trade, division of labor, varieties of products and services, economic opportunities, wealth, and support for the arts. Still more important, the growth of urban societies meant the assertion of a burghers' culture which, in opposition to the feudal tradition, placed production above fighting; the skills of reading, writing, and counting above the skills of tournament and hunting; practicality over pomp. From the 11th century onward it was this culture that on an increasing scale nurtured and molded the economic and technological development of Europe and eventually of the world.

The illustrations which follow suggest the importance of the Urban Revolution that took place within the walls of the medieval cities. But the Urban Revolution itself cannot be illustrated.

The walls and towers of Carcassonne in Languedoc, France confront the Pyrenees and for over a thousand years they were far more impenetrable. On defense foundations first fashioned by the Romans and the Visigoths, the French kings of the 13th century built up the supreme medieval fortified city: self-contained, secure, an independent world of its own, though ultimately depending on the agricultural production of the land beyond the walls. Intellectuals, goldsmiths, armorers and craftsmen, even actors engaged to play in an open-air theater, all worked within the city, which in its turn was dominated by the citadel of the administrators' château, a centerpoint from which to suppress town riots as well as the last stronghold in any siege. Wheat was brought in to the city mill in a surplus that allowed it to be stored. Beasts were slaughtered and salted. One room alone in the great Narbonne Tower could hold the carcasses of 1,000 pigs and 200 sides of beef. Long before its final fortification, which did in fact ensure that the urban shrine which came to be called the Virgin of Languedoc was never violated, the city had demonstrated its strength. After a five-year siege by Charlemagne a defiant citizen, Dame Carcas after whom the town was thenceforth named, took a pig to the top of the wall in full sight of the investing army and nonchalantly fed it with the last grains of the wheat stores as if there were countless bushels in reserve. The emperor abandoned the siege.

A massive order for Flemish bricks, given in Ypres in 1278 by Edward I of England for work on the Tower of London, marked the revival of brick building in England, where brick manufacture had been discontinued since the Romans left the country 800 years previously. Tiles, which were made by the same process, had been compulsory for roofing in London since 1212, when shingles and thatch were condemned as a fire hazard.

The specialists in brickmaking and bricklaying came from Flanders, which had used up its stock of building-timber earlier. Apart from the immigrant bricklayers and skilled freemasons working on stone, the master craftsmen of 13th-century building were carpenters, whose responsibility included building the framework to support a masonry arch until the keystone was dropped in. In the illustration from the *Book of St Alban* a carpenter is trimming a beam with an axe, and it was the carpenters, managing the primitive scaffolding of the day who set up the simple pulleys and windlasses of the hoists.

The horizontal loom perfected in Europe in the 13th century was a readoption of the style of weaving machine first used in the Middle East some 6,000 years earlier. As a groundframe it was common among the Egyptians from 3000 BC, but the only loom used by the Greeks and Romans was vertical, and the date of reversion is uncertain. The positive improvement was the provision of pedals lifting rods to which the warp threads were attached. In the simple two-pedal design alternate warp threads were guided by one of the two rods. When a pedal was depressed one rod was raised to form a *shed* through which the shuttle holding the weft thread was thrown from one hand to the other. The pedal was released, the weft was beaten into the body of the cloth, and the process was repeated with the other pedal. The lever on the weaver's right was operated to unwind more yarn from the warp beam and the woven cloth was wound up on the breast beam. The cloth, which appears vertical, was woven horizontally, but its aspect shows the narrow width usually woven. The horizontal loom was to remain static in its efficiency until the introduction of the flying shuttle in the 18th century.

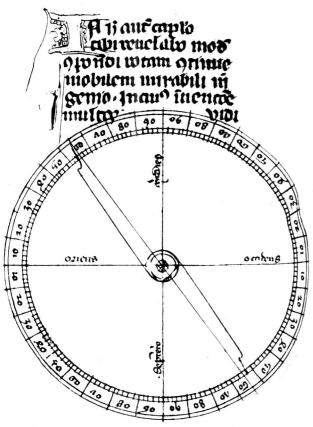

Accurate navigation was markedly improved in the 13th century by the accelerated adoption of the mariner's compass exploiting the magnetic needle, and by detailed charts based on careful use of the compass.

The properties of the lodestone (leading stone) had been known for two millennia.

This magnetic oxide of iron was recorded in Greek records of 800 BC as being mined in Magnesia, a province of Thessaly. The mineral was one of the sacred Chinese stones, and pieces of lodestone, cut into the shape of a spoon to represent the Plow or Big Dipper constellation which points to the north, were used for divination: the fortune-teller reserved to himself the knowledge that a lodestone carefully skimmed onto a smooth surface would always point the same way. The fact that iron touched by the lodestone became (briefly) magnetized was discovered later.

By 1100 AD the principle had been adopted in China to make the mariner's compass by fixing a strip of iron onto a disk of wood and floating it in a bowl of water. It was first noted in Europe in 1187 by the English sage Alexander Neckham. More valuable work on the compass and the deeper theory of magnetism was done by the French investigative scientist Pierre de Maricourt, Peter the Pilgrim. In his work *De Magnete* of 1269 he described and illustrated a new floating compass graduated with 90 degrees in each quadrant and adapted with movable sights for taking bearings. Using iron strips on a sphere of lodestone he mapped the magnetic field and, because the vertical lines intersected north and south like lines of longitude, he originated the term *magnetic pole*. Seeking to achieve perpetual motion, then a fashionable quest, he fitted a vertical needle with cups at its points, used a magnet to swing the needle up from south to north, and a runaway ball to carry it down from north to south and swing it into upward orbit again. His surmise that magnetism could be converted into kinetic energy was not to be practically demonstrated until Faraday's work 550 years later.

Consistent application of the mariner's compass led to the far more accurate charting of coastlines, gained by plotting bearings from point to point. These were published in a series of shipmaster's manuals named *portolani* originated in the 13th century by the Genovese and Pisans to describe the seaports and chart navigational bearings for the entire Mediterranean coast. The charts in a portolano gave a network of bearings radiating from each port. The *Carta Pisana*, a sheepskin chart of 1275, is the earliest surviving portolano chart, depicting the Mediterranean from the Straits of Gibraltar to the Holy Land.

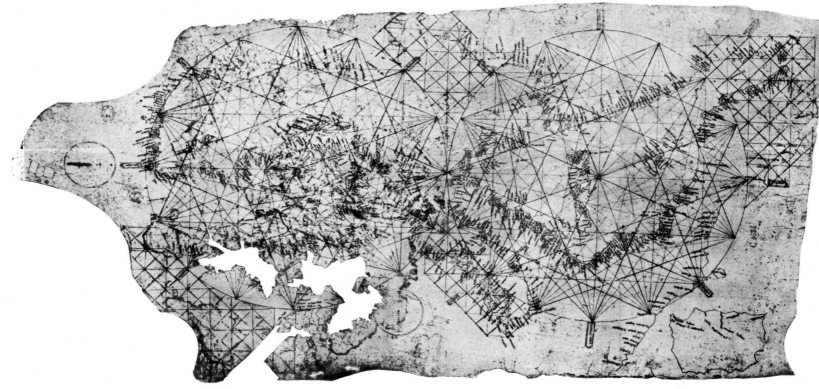

The city walls of Zurich march down the slope of the shore and advance into the water, with a guarded narrow watergate emphasizing the security and community of interest within. Carts bring grain to be ground in the watermill, established as a civic monopoly and recognized as a bastion of communal subsistence. But beyond its assurance of parochial cosiness the city exerted power on a continental scale. Zurich, originally a church fief, straddled the European trade routes, and its busy merchants had won autonomous privileges by 1000 AD, and complete independence from church suzerainty in the 12th century. In 1218 it became an imperial free city, giving its corporation the status of any prince of the Holy Roman Empire.

THE WATER MILL AND THE WINDMILL

For all their creativity and efforts, pre-industrial societies operated under one severe constraint. Perhaps 80 to 90 percent of the total energy consumed at any one time before the Industrial Revolution was derived from plants, animals, and men.

This figure set an intractable limit on the productivity of pre-industrial societies and on the range of tools and machines they could produce and use. Once this is understood, we can appreciate the importance of the sailboat, the water mill, and the windmill – three energy converters which made it possible for pre-industrial man to exploit the inanimate energy of the wind and of water streams. To the extent that man could use one or another of these converters, he could break through the energy bottleneck that trapped him and limited his productivity. The success of the most powerful societies of the past, from the Phoenicians to the British of the 17th century, was based on their extensive exploitation of the energy of the wind on the seas.

Water mills had been in existence both in the West and in the East since at least the 1st century BC, but it was in the primitive Europe of the Dark Ages that the water mill began to emerge as an important device. This is in itself astonishing, because the mechanization of productive processes is not normally led by the most primitive areas. Some historians have offered a

Overleaf: Watermill with women grinding grain, from *Hortus deliciarum*, 12th century

simple explanation: classical antiquity had an abundance of slaves who provided an abundant supply of alternative energy; the decline of slavery and the demographic depression forced the lords of the Dark Ages to have more ample recourse to the mill.

Closer scrutiny, however, shows that reasoning to be too simplistic. In the Dark Ages some monastic communities reached a remarkable size. The Abbey of St. Riguier under Angilbert (833 AD) numbered 400 monks; and the Abbey of Corbie under Adalhard (822 AD) sustained about 400 people within the monastic premises, as well as some 400 knights. The problem in such cases was not so much the numerical ratio between the lords and the serfs but the discrepancy between the concentration of the lords in one place and the dispersal of the serf over the estates. This, however, was not the whole story.

As feudalism consolidated, the feudal lords had the power to make their estate mills profitable monopolies. They not only made their mills available to the local peasants in return for payment; they also sent out their manorial agents to seek and destroy the peasants' hand mills so as to oblige the peasantry to bring grain to the lord's mill. Such ruthless monopolistic practices swelled estate revenues and undoubtedly favored the spread of a technology which had been known for centuries.

While the proliferation of the water mill was in full swing, another innovation arrived to permit the harnessing of more inanimate energy. Prototypes of windmills were probably known in Persia as early as the 7th century AD, but toward the end of the 12th century a totally new type of windmill appeared in Europe. With its sails mounted on a horizontal axis, the European windmill had very little in common with the Oriental mill (whose sails were mounted on a vertical axis) and was probably a totally separate invention. The new machine first appeared in Normandy and in England; later, windmills are mentioned in 1204 in Picardy, in 1237 in Tuscany, in 1259 in Denmark, in 1269 in Burgundy and in 1274 in Holland. Originally, the Western windmill was mounted on a post and the whole mill had to be turned to face the wind. By the 14th century, however, the tower mill had been developed: the building and the machinery remains stationary and only the top rotates to face the sails into the wind. Thus tower mills could be built much larger than the post mills, to produce as much as 20 to 30 hp.

To understand the saga of the mills in the first centuries of our millennium we must, however, leave the history of technology proper and venture into social and economic history. As we have seen, cities grew in number and in size from the 10th century onward and this urban development was accompanied by a sustained expansion of manufacturing production. As trade and manufacturing grew, the motive power derived from hydraulic and wind energy was applied to an increasing variety of productive processes. Thanks to the adoption of mechanisms moved by cams set on the axis of the mills, hydraulic power was used for the fulling of cloth in France around the year 1000, in Sweden in 1161, and in England by 1185. There is evidence of the use of mills in the production of iron in Styria in 1135, in Normandy in 1204, and in Southern Sweden in 1224. In Normandy a saw mill was operating in 1204. Mills were being used in the manufacture of paper at Fabriano in 1276, at Troyes in 1338, and at Nuremberg in 1390.

At about the same time, mills ceased to stand only in the countryside and began to proliferate within the walls of cities as well. In the mid-16th century a traveler describing the city of Bologna wrote that within the walls a sluice from the river Reno turned *various machines to grind grain, to make copper pots and weapons of war, to pound herbs as well as oakgalls for dying, to spin silk, polish arms, sharpen instruments, saw planks.* Windmills could not be built within city walls but, as at Amsterdam in the 17th century, they could be built along the ramparts around the city – not only for milling grains but for other manufacturing purposes as well. Thus mills became instrumental in the mechanization of a number of productive processes, and manufacturing activities typically developed where water power was available. One example may illustrate the importance of these developments.

Paper had first appeared in China some time before 100 BC. The Arabs learned the technology from the Chinese around the middle of the 8th century AD. Both the Chinese and the Arabs in the East continued to hand-make paper. But when the production of paper spread to the West, water power was used immediately. At Fabriano (Italy), which was the main Western center of paper production in the 13th century, water mills were extensively employed. Mechanization meant reduction of costs and therefore of prices. So, from the end of the 13th century the paper from Western mills began to compete successfully in Middle Eastern markets with the local product made entirely by hand. It eventually brought ruin to the local manufacturers.

It was an early, though inconspicuous, announcement of the Industrial Revolution of five centuries later.

Water mills were used in China for industrial purposes beyond their traditional function of making flour for about 1,000 years before their adaptation in the West. But by the 11th century their use was broadening in Europe.

It was water power that particularly excited the French Cistercian architect and engineer Villard de Honnecourt, who in about 1240 compiled on parchment a notebook with rough sketches, hoping that his fellow professionals *would find it useful in building, in constructing machines, and in the application of geometry to the plotting of figures.* Villard's own figure plotting was somewhat ungeometrical, and essential sections are missing in certain of his designs. His most important discovery of the application of water power was linked with the hydraulic saw shown to the left of his not very original crossbow.

A water-driven saw was mentioned in a suspect manuscript of 369 as having been operated to cut marble on a tributary of the river Moselle. Villard's authentic description occurs during an era when water power was known to be driving fulling mills for thickening woolen cloth, trip hammers, and iron forges, as well as the traditional grinding of grain.

Villard's water-powered saw used a downflow of water to rotate a horizontal wheel working a cam which pulled the saw down. The upstroke of the saw was accomplished by the spring of a sapling which had been bent as the saw came down – a variant of the spring principle of the pole lathe, freeing both hands of the wood turner, which was introduced at about this time. Villard seems to have omitted a mechanism which drove the workpiece forward against the saw.

His water-driven pile saw, to cut off poles driven into a river bed so that a bridge or jetty could be built on top, is also incomplete. Between the two hydraulic machines he illustrates part of a device for mounting a wheel on an axle. Beneath the crossbow is a screw jack, as simple as a modern car jack, which astonished Villard as *one of the strongest appliances in existence for lifting weights.* Next to it is a clock-style escapement used for rotating the figure of an angel on a church roof, a not uncommon medieval "wonder." Villard, like Hero of Alexandria, was much taken with automata toys, and his articulated eagle, with a ring-fitted head, was made to stand in church and turn its face toward the deacon as soon as he began to read the Gospel.

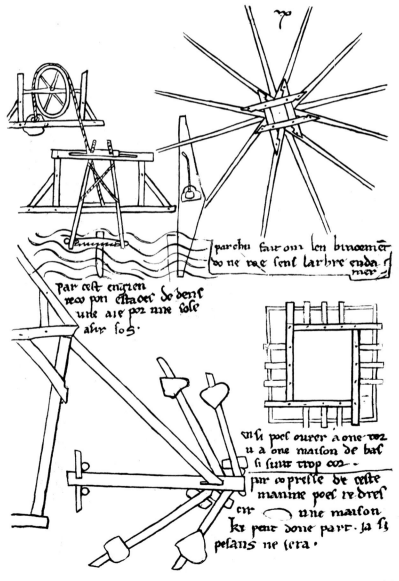

Water mills were mentioned in Egypt, Thessalonica, the Black Sea coast of Asia Minor, Jutland in northern Denmark, and the far shores of China, all in the century between 65 BC and AD 31. *The almost simultaneous appearance of this first power machine in regions as widely separated*, suggests Professor Lynn White ... *argues diffusion from some still unknown centre, presumably north and east of the Roman Empire.* A horizontal wheel, cheap to build, needed a constant flow of fast water, and had little power. Turning slowly, and with no gearing to make the millstone turn any faster, it would grind sufficient flour for one family. An industrial water wheel of far greater efficiency was designed by a Roman engineer in the 1st century BC, and was described, with many other invaluable references, by Marcus Vitruvius in his work *De Architectura*.
The revolutionary element in the power source, now called the Vitruvian mill, was that the wheel was vertical and it was geared to drive the horizontal millstone at a speed five times greater than its own revolution. It began as an undershot wheel, with the flow of water working on the blades at the bottom (the exact reverse of the powered paddle wheel).

Some centuries later – the first known example was installed in the marketplace at Athens in 470 – the overshot wheel was introduced. Here, the movement and weight of the water is applied to the blades or buckets in the upper part of the wheel. Greater power is generated, but more ingenuity was needed to collect the water. It had to be stored in a millpond, using a dam or weir, and delivered to the millrace by a cut channel or sluice. The bed of the stream had to be cut to accommodate the entire depth of the wheel, and the chute controlling the angle of fall had to be designed correctly.
The stream illustrated in the Luttrell Psalter of 1338, dammed to raise water for the overshot wheel, has traps for eel and other fish set in the race. The adoption of the Vitruvian vertical wheel caused design problems which ultimately benefitted all mechanical adaptation. Gears had been an invention of the Hellenistic era, but before the water wheel little had been required of them except to drive a water clock or the low-powered AntiKythera computer. Now they had to endure great strain, at ratios producing good efficiency, for long spells at high speed. This pragmatic field-testing occupied centuries. The float mill – a wheel operating between two boats moored in a fast stream – was improvised in 537 by the Byzantine general Belisarius defending Rome against the Goths. A tidal mill was working in Dover harbor before 1086, when it was recorded in the Domesday Book.

In terms of technological importance, the most significant use of the water mill was its incorporation by the Italians into silk manufacture. When silk is thrown, the filament is continuous, not intermittent as in vegetable fiber. Water-powered, silk-twisting mills were set up in Bologna in 1272.

Until the introduction of the steam engine, water mills and windmills were the sole sources of industrial power. Though the principle of windmills had been mentioned three centuries earlier, their first systematized use was reported in 950 from the Persian province of Seistan, where they were used for corn grinding and irrigation pumping. Later they were adopted in Afghanistan, and by the 12th century had spread eastwards to China and westwards throughout Islam.

The wheel carrying the sails was originally horizontal, like the early water wheel, and similarly inefficient, transmitting power during only half of one sail's revolution. Western Europe utilized from the start an adaptation of the Vitruvian water wheel: four vertical sails driving a horizontal shaft. The first windmills in western Europe were post mills, where the sails and machinery were mounted on a stout post, and the entire apparatus had to be rotated to face the wind. The earliest reference to a post mill is in Normandy in 1180.

Two centuries later the tower mill had been introduced, enclosing the machinery in a stationary tower so that only the cap carrying the sails needed to be turned to the wind. Such a tower mill features in a stained glass window of Stoke-by-Clare church, c. 1470.

At first, Western windmills were used solely for grinding corn, but the Dutch applied them to drainage in 1430 and to timber sawing in 1592. Power was progressively increased, developing in the 16th century a capacity equivalent, as rated at that time, to three water wheels, 25 horses, or 300 men. The fantail, regulating the mill to keep it facing the wind by the first application of automatic control, was to be a late British adaptation introduced by Edmund Lee in 1745.

A spinning helicopter toy, illustrated in the early 14th century but used by children throughout the ages, emphasizes that the helicopter is the first aircraft type to have been demonstrated as a practical possibility, whilst being the last to be realized. The toy continues today, working on the same principle as is shown in the Flemish psalter, excusing the illustrator's technique in perspective which shows a horizontal lifting surface as a vertical windmill-sail. A freely bedded horizontal airscrew is whirled round fast on a corded spinning-top until it disengages from its mounting and lifts into the air.

The spinning of wool, and the fibers of other textiles, was given only two technological improvements in its first 7,000 years of practice. The routine of rolling the fibers between the hand and the thigh was supplanted by the use of the spindle, a wooden core holding the carded fiber which was kept rotating by the human hand, twirling the fibers as they were drawn and spun into yarn. Over 5,000 years after the introduction of this mechanical aid came the innovation of the spinning wheel. The action is accurately shown in this 1338 illumination, but the spindle itself, rotating in the same plane as the wheel, is not clear. The spinster pushed the wheel round by the right hand. The wheel rotated a spindle by means of a belt and pulley. The turning spindle allowed yarn to slip from it as the operator drew it with her left hand. The spinster, who started the sequence leaning forward, is shown at the end of her throw: she cannot draw back any more without losing her rhythm at the wheel. At this point she has to pause and wind the yarn. The spinning wheel was later improved by a device called the flyer which allowed spinning and winding to be done simultaneously.

This continuous rotary motion of spindle and flyer was estimated by Adam Smith to have doubled the productivity of spinning labor. There was to be no parallel speeding of the weaving process until all textile workers endured the revolutionary shocks of mechanization in the 18th century; the spinning jenny and mule, the flying shuttle, waterframe, powerloom, and the carding-engine. Carding, the disentangling of washed wool before it was fit to be massed on the spindle, is being done by the woman on the right, using a brace of wire-toothed brushes, a device then about a century old.

THE MASTERS OF BOMBARDS AND CLOCKS

Overleaf: Portrait by Tommasso da Modena, showing spectacles, 1362

In the summer of 1338 a galley left Venice bound for the East. The galley carried, among other things, a clock – probably a mechanical clock – that Giovanni Loredan was hoping to sell in Delhi. We know of the event because some merchants later pursued claims about the cargo in a court of law. No contemporary chronicler noted the event and modern historians scarcely mention it. Yet that was a fateful shipment: Europe had begun to export machinery to Asia. A new era was unfolding. The age of European under-development was manifestly over.

Sundials and clepsydras had been the principal instruments used for the measurement of time from earliest antiquity. In 1271, in his commentary on *The Sphere of Sacrobosco*, Robertus Anglicus mentioned that *artificers are trying to make a wheel which will pass through one complete revolution for every one of the equinoctial circle, but they cannot quite perfect their work*. In 1309 a mechanical clock made of iron showed the time on the bell tower of the Church of St. Eustorgio in Milan. So, some time between 1271 and 1309 the mechanical clock, as we know it, was invented.

The first mechanical clocks were hardly prototypes of the precision machine. In fact, medieval clocks kept such poor time that they had to be continually reset with the aid of the traditional timekeepers, the sundials and the clepsydras. Their adoption was more a reflection of the fascination of Western society with the machine than a function of the efficiency of the machines themselves. Despite the fact that the new contraption kept very poor time and had to be continually adjusted, communities went to great effort and expense to have – as a 15th-century French document put it – *a great and honorable clock which would give prestige to the town*. This fascination with the machine had far-reaching effects.

It infused philosophical thought, and the mechanistic outlook came to dominate every branch of human knowledge. A few centuries after the masters of bombards and clocks had first produced their rudimentary contraptions, Kepler was to assert *the universe is not similar to a divine living being, but is like to a clock*. And Descartes wrote: *I recognize no difference between the machines made by craftsmen and the diverse bodies put together by nature alone*.

We marvel at the progress of technology because we generally look only at those tools and machines man has used to overcome the necessities and miseries of life. But there is also a dark side of the story which demonstrates that technology is not a blessing in itself.

This book contains no illustrations of instruments of torture, nor does it contain more than a few illustrations to show the development of weaponry. However, the pictures of early firearms which follow are essential.

Though the Chinese probably invented gunpowder, they are believed to have used it only for making fireworks. In Western Europe man applied the properties of gunpowder to artillery. An official document of 1326 refers to the acquisition by the Florentine government of bronze guns shooting iron balls. Some 30 years later Petrarch wrote: *these instruments, which discharge balls of metal with most tremendous noise and flashes of fire, were very rare a few years ago and were viewed with great astonishment ; now they are as common and familiar as any other kind of arms, so quick and ingenious are the minds of men in learning the most pernicious arts.*

The date of the first cannon is certainly not fortuitous. Between the mid-13th century and the mid-14th, Europeans invented (among other things) the spinning wheel, spectacles, and the mechanical clock. It was a period of exceptional creativity. There was little scope for specialization and many masters shifted indifferently from the production of clocks to the production of bombards (artillery), and thus were known as *magistri bombardarum et horologiorum* (masters of bombards and of clocks).

The development of artillery led to vital consequences in various fields of technology and science as well as in economic and social affairs. The new invention slowly but irresistibly forced drastic changes in the art of war, in the art of fortification and other related architectural problems, in the development of mining and metal production, and in the development of applied mathematics and the science of ballistics. It hastened the military and social decline of the knight and it encouraged the formation of larger political states.

At first, firearms were mostly used by Europeans against each other, and so the original invention was quickly and continuously improved. Paradoxically, this put Europeans at a relative advantage over the rest of the world, insofar as artillery became an important instrument of the overseas expansion of Europe.

To primitive man the discovery of fire was essential for survival. Later civilizations used the destructive as well as the conserving force of fire. Burning arrows and lances and fire pots were used as projectiles. In the Middle East the Byzantines and Moslems made incendiary projectiles from available minerals such as sulphur, saltpeter, and petroleum compounds with low flash points. *Greek fire* was a liquid incendiary often projected by a force pump from the prow of a ship.

Beginning in the 13th century natural combustion was manipulated – probably first in the Middle East and later in the Far East – to produce a semicontrolled explosion made possible by gunpowder. The first Western representation of gunpowder occurs in an illustrated manuscript completed in 1327 under the supervision of Walter de Milemete, who wrote the text and dedicated it to his pupil, the 14-year-old British King Edward III.

This manuscript, *De Officiis Regum*, defining the duties and powers of kingship, includes an illustration of a windsock kite being flown over a besieged city and carrying an explosive or incendiary bomb, to be released by cutting the cord.

It is unlikely that the illustrator had ever seen the original bomb or the gun. Milemete does not mention the weapons, though there is a textual reference to *metal cannon shooting iron balls or darts* from Florence in 1326.

The graphic details are probably inaccurate. Certainly the wings which have been added to the ancient windsock draco are far too small in the picture to achieve their aerodynamic purpose. The gun, which is being touched off by a soldier carrying a red-hot iron rod, is projecting a four-headed metal dart, based on the design of the crossbow quarrel. From the shape of the original gun the first artillery pieces were called *pots* by the French and *vasi* (vases) by the Italians, but the more cylindrical barrel – describing a method of construction from iron staves hooped together – was adopted within a few years.

In 1344 Jacopo de'Dondi, a 51-year-old physician and astronomer, made a clock to be set in the Carrace Tower of his native Padua. He thereby earned lasting fame for himself and for his family the alternative surname *dell'Orologio* (of the Clock). In his memorial tablet on the wall of the Baptistery of Padua Cathedral Jacopo's epitaph emphasizes that the passerby *could from the top of a tall tower tell the time and the hours, though their number changes.* Nine years earlier the first known striking clock bell (numbering the hours rather than indicating, as previously, the passing of an hour by one stroke) had been installed in a public clock in Milan. As yet, however, no clock dial had been recorded. The horologist H.Alan Lloyd suggests that the great innovation which earned Jacopo his honorary title was the introduction of the clock dial, an outstanding feature of the unique clock constructed in 1364 by Jacopo's son Giovanni.

Giovanni de' Dondi, born in 1318 when his 25-year-old father was municipal physician at Chioggia, became professor of medicine, logic, and astronomy at Florence and Padua. In 1364, five years after his father's death, he completed a clock which became far more famous even than Jacopo's. He wrote specifications of it which were detailed enough for a replica to be made in 1960. Philippe de Maizières, who admired the original, recorded: *This master John of the Clock . . . has made an instrument, called by some a sphere or a clock for celestial movements . . . constructed in such a subtle way that in spite of the fact that there are so many wheels that they cannot be counted without taking the clock to pieces, all goes with one weight. So great is the marvel that great astronomers come from distant places to admire the work. In order to have his sphere done as he had it in his subtle mind, the said master John did nothing else for sixteen years.*

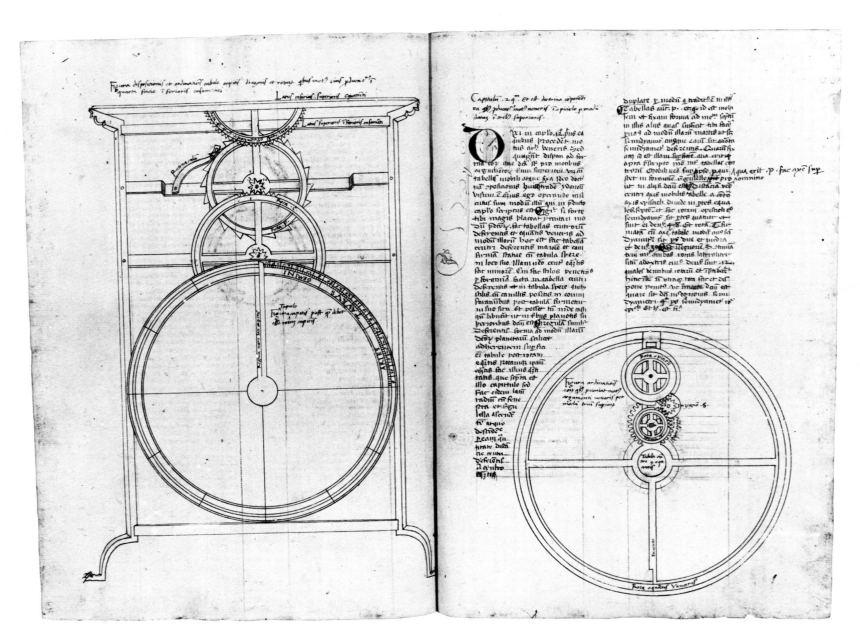

Giovanni's clock had eight dials, seven in a turret and a master dial controlling them, not to speak of the broad horizontal wheel of the governing universal calendar. The master dial showed the mean hour and minute on a 24-hour revolution. Mean time is the mathematical division of the *average* day into 24 hours of 60 minutes. Because of the Earth's eccentric orbit round the Sun and its slanting axis, the longest solar day in the year is half an hour longer than the shortest solar day. The ultimate standard of time, a sidereal day, is the duration of one rotation of the Earth relative to the stars. A solar day is the duration of one rotation of the Earth relative to the Sun.

John the Clock's mean-time dial was geared to show sidereal time on another dial, and appendages showed the times of the rising and setting of the sun. The sidereal-time mechanism drove a dial showing the cycles of the Moon. Separate dials driven by the annual calendar wheel showed the position of the planets through the year (five planets were then recognized). From various other indications of the great clock there could be read the temporal hours (the monastic division of daylight and darkness into 12 equal parts varying greatly from season to season); the dominical letter identifying Sundays from year to year; and both the fixed and movable feasts of the Church. His perpetual calendar for ascertaining the date of Easter was not paralleled for nearly 500 years.

Clocks became status symbols in cities, ringing out the prestige of 14th-century burghers. Bellfounders spread out from northern Italy and the Rhine basin. In 1368, three horologists were given passports to travel from Delft to England to practice their craft for a year. Salisbury Cathedral boasts the oldest existing weight-driven mechanical clock known, dating at least from 1386. It was originally housed in a bell tower in the close but was placed inside the cathedral in the 18th century. It strikes the hours but has no dial. After 1844 it was abandoned, but it was found in 1929 by the horologist T.R.Robinson who, after X-raying the mechanism, removed accretions and fitted reconstructed parts on the original pattern. The clock was re-established as a working instrument in the west side of the nave.

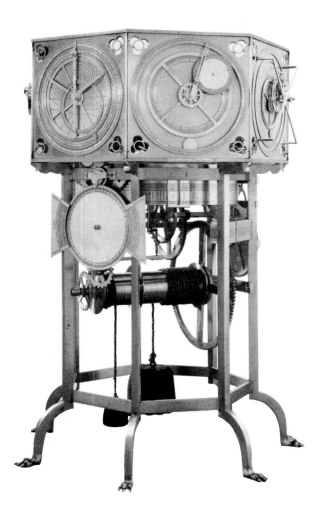

The *pot-de-fer* shown shooting the quarrel in 1327 was probably made of cast bronze, an alloy of copper and tin, and the European craftsmen who met the surging demand for bronze church bells were soon just as handy at casting cannon. Wrought iron was a cheaper alternative, but less favored by the military, who preferred not to die from their own burst breeches. The shot consisted of balls of either iron or stone. Within 20 years cannon were in high esteem and the era of mass slaughter by ballistic missiles had begun, not very efficiently at first, with the Hundred Years War. Edward III took cannon to Crécy in 1346, but their main effect seems to have been to frighten the French horses, and the battle was decided by the English longbow men.

At the siege of Calais, begun in the same year, at least 20 cannon (*bombards*) were used. A contemporary jingle asserted:

> *Gonners to schew ther arte*
> *in to the towne in many a parte*
> *schote many a fulle gret stone.*
> *Thankyd by God and Mary myld*
> *They hurt nothir man, woman ne child.*
> *To the housis thow they did harm.*

The siege guns were not decisive at Calais. The siege lasted 11 months and the city finally capitulated through starvation. There still existed far more fatal scourges than artillery. At the time of Crécy the plague of the Black Death was raging in China. During the siege of Calais it ravaged India and the Middle East. Within two months of the fall of Calais it had reached Messina, Sicily. Gradually it spread across continental Europe. Throughout the world which it hit, the plague killed one in three, a million and a half in England alone. It was all over by 1350. It was not the Black Death but the Black Prince who beat the French at Poitiers. Edward III contemptuously sent King David II back to Scotland after 11 years' imprisonment in England, and as a precaution invested in more artillery. Cannon quickly became huge. Mons Meg, constructed in Belgium in the 15th century and now at Edinburgh Castle, weighs five tons, has a caliber of 20 inches, and fired a stone ball weighing 330 pounds over 2,500 yards.

The new demands of gunmaking expanded certain handicrafts at a time when the craft guilds were at the top of their prosperity and ripe to exploit new techniques. The gunsmith, the powder maker, and the military engineer acquired a new importance – often their skills were mastered by the same person. The trade of ironfounder, begun in the Rhineland early in the 14th century, prospered. In military technology the Germans became ascendant. *As Heaven is adorned with stars, so Germany shines by her free craft and takes honor for her mechanical knowledge,* wrote the military engineer Konrad Kyeser von Eichstätt in his manual *Bellifortis* in 1405.

A master gunsmith needed skill and new technology, not only to cast, forge, and work metal for artillery but also to shoot with cannon and survive the manufacture of gunpowder. Making gunpowder was always dangerous. The ingredients were sulphur, charcoal, and saltpeter, originally mixed in proportions of 2 to 1 to 6 by weight (the more successful modern recipe is 1 to $1\frac{1}{2}$ to $7\frac{1}{2}$); they first had to be pounded with a pestle and mortar, though later the work was done with water-driven mills. In gunpowder stamps of 1470 two-handed vertical pestles were mounted on sapling boughs to enable them to spring up after each blow. The mixture was kept damp while it was being worked, but all too often explosions occurred.

In a medieval illumination, several ballistic weapons – longbow, crossbow, bombard, and handgun – are illustrated with surrealist perspective. The most surprising feature is the length of the handgun and the expertise with which it is handled. The gun had to be fired by advancing the right hand, which carried a coal, towards a touchhole at the top of the tube; this operation took the shooter's eye off his target so that at first it was carried out by another man. The earliest handguns were a spin-off from the antipersonnel cannon called the *ribauldequin*, first chronicled at Bruges in 1339. This was a frame supporting a number of thin tubular barrels – 144 barrels were ambitiously mounted by the Lord of Verona in 1387 – which could be fired almost simultaneously with the sweep of one coal. *Ribauldequins* were mainly used for defending castle entry corridors against besiegers who had already stormed the gates. Within 10 years of their inception their dismantled tube barrels were being used separately as handguns. A typical length used in 1364 was a 12-inch barrel. By 1374 the English were mounting these barrels on wooden stocks four or five feet long. Gradually the barrels were lengthened and the stocks shortened. The match, a slow-burning twist of rope, was substituted for the coal, and made the gunner independent of a brazier. By 1425 a lock had been introduced, mechanically applying the match to the powder.

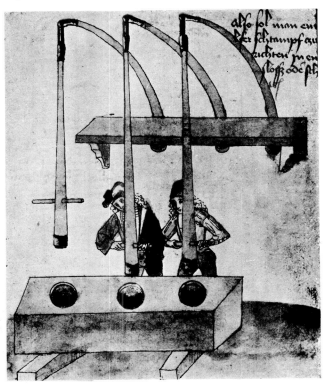

The plow began to be drawn by oxen in Sumeria, Babylon, about 3000 BC. It reached central Europe, via southern Russia, around 1100 BC, by which time the spread of iron technology after the fall of the Hittite empire was making the use of iron plowshares more common. In its first evolution from the digging stick it had the form of a scratch plow with a conical share, not effectively turning over the soil but leaving untouched margins which could be broken up only by cross-plowing. This was no great handicap in the dusty soil of western Asia and the Mediterranean basin. But the heavier soils of northern Europe demanded stronger power – often a team of eight oxen – and a new form of heavy plow. It had a vertical knife as the coulter, a horizontal knife as the share to cut at the level of the grassroots, and a moldboard which turned the turf over the side of the furrow. This efficient instrument, introduced in near-perfection to Britain and Normandy by the Danes in the late 9th century, made cross-plowing unnecessary and encouraged strip farming in large open fields. It further affected rural life by making larger communities (with their social intercourse and necessary policy-making institutions) more favorable units than scattered hamlets. Even the beasts in an eight-ox team had to be contributed in ones and twos by individual peasants, who in return owned or managed scattered strips in the plow land.

By this time the stronger horse was being bred for military purposes and being fitted with iron shoes. The horse was increasingly used for traction. But the old system of harness, which suited the anatomy of the ox, used neck and body girths which cut into the windpipe and the jugular vein of the horse, while the yoke, bearing on the withers between the shoulder blades, was inefficient for traction. A new form of harness was devised and was widely used by the late 12th century. It utilized a padded collar to which the traces were attached, so that the draft effort was taken by the horse's shoulders. A horse so harnessed could pull between four and five times the weight of a yoked horse, and had greater speed and longer endurance than the ox. Using this extra power, the men of the Middle Ages built longer carts. The horse pulling a long cart and the obsolescent yoked oxen were both illustrated in the Luttrell Psalter of about 1338.

The *Mendelschen Zwölf-Brüder-Stiftung* issued in Nuremberg around 1390–1400 shows workmen using instruments which had changed little in a thousand years, though the date of publication coincides with the first historical mention of an iron founder, Mercklin Gast of Frankfurt-am-Main.

The turner works a pole lathe of the type used by the earliest forest craftsmen who kept a live sapling growing over their hut. The workpiece is mounted in the frame horizontally. A cord passes from a treadle, around the spindle, and up to a springy pole. On the downstroke of the treadle the work revolves a number of times against the cutting tool. As the pole draws the cord back, the work revolves the other way and the chisel is used more lightly to clear the shavings.

The same principle of alternate revolution is used by the workman boring beads for necklaces. His bow-string drill (with the string encircling the chuck) is of a type believed to have been used in the eastern Mediterranean 3,000 years earlier.

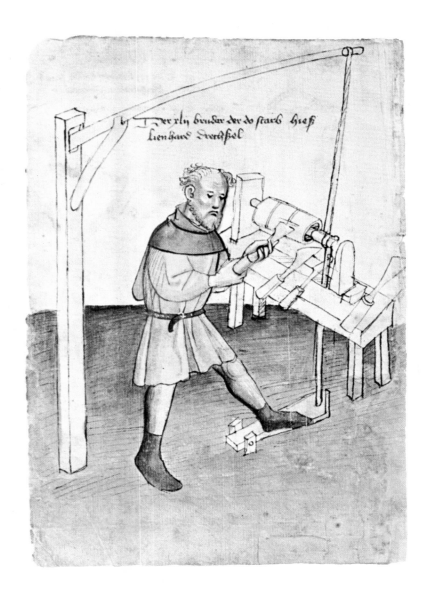

In his treatise on military technology, *Bellifortis*, written in 1405, Konrad Kyeser mentioned a *draco volans*, or flying dragon. This indicates the advance of the soldier's draco from a military pennant to an aerodynamic kite. As early as 105 AD the Romans had used a *draco* as regimental colors, as a signalling medium, and as a windsock enabling archers to judge the wind. It was retained in Europe for 1,500 years, gradually evolving into a kite.

By the Middle Ages the windsock pennant was often fitted with a combustible which made fire gush from its mouth and smoke stream from its tail. Kyeser indicates that by 1405 the windsock was being tested as an offensive pyrotechnic kite, necessitating plane-surface wings (which are only inadequately shown in Milemete's manuscript of 1327 and entirely omitted in the *Bellifortis* illumination). The instructions for making it suggest that the body may be a flat two-dimensional structure of linen and silk (making it a true kite with an aerodynamic surface) with a separate three-dimensional head braced by the wooden triangle to which the cord is attached. Kyeser did not specify how the implement was to fulfill its purpose of projecting fire into an enemy camp. Within a generation plane-surface pennon kites were replacing the heavier windsocks.

An entirely different application was attempted, probably around 1420, by Giovanni da Fontana. Rockets may have been known to the Byzantines by the year 1000, and were later used by the Chinese for fireworks and by the Saracens for projectiles. Fontana toyed with using their reactive force for practical research. He devised a mechanical rocket-propelled bird which flew, a fish which dived and swam, and a rabbit which ran forward on rollers – all illustrated in his sketch – as means of measuring heights, distances, and depths of water. If his jet-propulsion bird ever did fly it was a pioneer aerodyne.

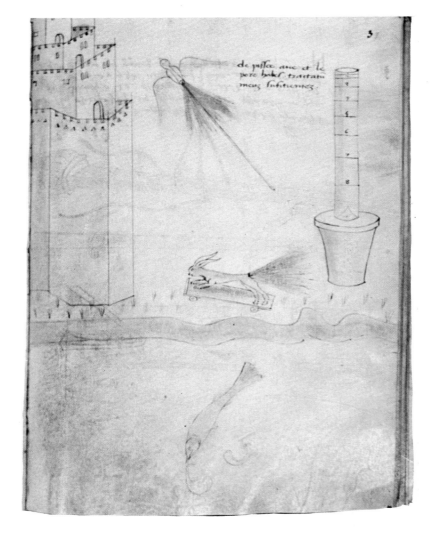

THE PRINTING PRESS

At the end of the 8th century in Spain a book cost roughly as much as two cows. In the 14th century in Northern Italy, a book on medicine could cost more than half an ounce of gold and a law book more than three-and-a-half ounces of gold. Private libraries (as against the libraries of princes or abbeys) rarely contained more than 10 to 50 books; but small as they were, they represented great fortunes. Their owners always took great care to ensure that these treasures should be passed on to the right hands. Few people could afford to buy books, and students learned not from books but by listening to lectures and taking notes.

Nor did the introduction of paper in the West after the mid-13th century help, for by far the largest item in the production cost of a book was the wage of the copyist: it was a waste of money to hire a copyist to write on such a relatively perishable material as paper. However, the spread of paper mills in the West was important in paving the way for the invention of printing from movable type.

Like paper, printing was a Chinese invention. And as in the case of paper production, the West borrowed the initial idea of printing from the East but modified it to make it suitable for mass production. For the basic innovation in printing from movable type was a dramatic cut in production cost.

Overleaf: first known illustration of a printing press from *Danse Macabre* published by Mathias Huss in Lyons, 1499

110

The new system was remarkable for its rapid spread. The first book printed from movable type appeared in 1445 at Mainz in what is now Germany. Within 30 years or so the new technology had reached north to Oxford, south to Palermo, west to Valencia, and east to Budapest. Within the following century it had spread to China, India, and the Americas.

The first printers could print about 300 pages a day. By the end of the 15th century, the average had risen to over 400. By the beginning of the 18th century, two printers could pull about 2,500 pages a day. The still majestic *incunabula* of the 15th century were followed by the less beautiful but cheaper editions of the 16th century and then, in the 17th century, by small paperbacks – calendars, hagiographies, popular love stories and treatises of petty technology to be sold to a broad audience.

The public recognized the revolutionary importance of the new printing technologies. In the 19th century when Charles Babbage, the inventor of the first true calculating machine, wrote that *the modern world commences with the printing press* he was merely echoing the many 16th- and 17th-century writers who ranked printing, together with geographical discovery, among the main heralds of their new age.

Printing is a classic example of a technology whose development had a tremendous impact on society : it is also a classic example of the importance of social forces in either fostering or hindering the consequences of technological advance. Printing was invented at a time when the written word was slowly but steadily gaining ground all over Europe. The production and diffusion of relatively inexpensive printed books (and particularly of the printed Bible) contributed largely to the further progress of literacy in Western Europe after 1500. However, three centuries after the appearance of printing, illiterates still accounted for some 35 percent of the adult population in England, about 45 percent in Belgium and France, and about 75 percent in Spain and Italy. Social and economic structures and circumstances were the stumbling blocks which prevented the mass of the population from taking full advantage of a well-established technological advance.

Any technology offers potentialities to a society. But whether and how the potentialities are exploited ultimately depends on the values and structures of that society. The printing press helped spread both rationalism and dogmatism, both science and religion, both Harvey's ideas about the circulation of the blood and Galen's traditional fallacies about bloodletting. It helped circulate ideas among the cultivated, and multiplied the number of literate charlatans. Even so eminently civilized a technology as the printing press is ethically neutral.

After the publication in China in 953 of the Confucian Classics, printed from page blocks of wood, relief plates of copper began to be used for printing. Between 1041 and 1084 a Chinese artisan named Pi Shêng developed movable type. He cut characters in clay and baked them hard. He composed a page by assembling the characters within a frame of the required size. This chase was set on an iron plate covered with a waxy resin. He warmed the plate and pressed the frame of type into the soft resin with a flat board so that the type surface was absolutely flat, constituting the form. He then let the resin harden. The result was a block of type capable of printing identical pages. At the end of the printing operation he warmed the bed of wax, let the type fall out, cleaned it, and restocked it in the cases of his magazine. Later, in China movable type was cast from tin. Alternately, a large block of wood was carved with identical characters from a calligrapher's model, and was then finely sawed into individual rectangles.

After 1215, when Korea emerged as a center of culture, it led the world in printing, and developed the use of movable metal type. Specimens of such bronze type are disputedly dated from the 13th and 14th centuries. Certainly a central government typecasting foundry was set up in Korea in 1403, when the king Yung-lo ordained: *The books printed from blocks are often imperfect . . . I therefore order that characters be formed of bronze and that everything without exception on which I can lay my hands be printed, in order to pass on the tradition of what those works contain.* Enthusiastically, the Koreans set calligraphers to design beautiful type faces, and commandeered monastery bells and civil-service furniture to be melted down for the necessary type metal. But, though the Koreans had evolved a phonetic alphabet by 1440 (the date when Gutenberg started operating in Europe), they did not progress to casting single-letter type, and continued to make each piece of type represent one of many thousands of words, like the Chinese logograph.

Meanwhile, block printing, a process which was nearly 1,000 years old by 1440, continued to be practiced. It began as the printing of textiles and progressed to printing on paper. Old textile prints probably dating from the late 7th century were found in one of the Caves of the Thousand Buddhas in the region where the printed Diamond Sūtra was discovered. Japanese textile prints incontestably dated at 734 and 740 exist, and they contain the earliest examples of block-printed script. Within a few years, script was being block-printed on paper. Block-printed textiles began to appear in Egypt and western Europe shortly after this time, and it is likely that the technique of printing on fabric spread slowly from the Far East. The craft became much more widely practiced in Europe about 1400. At the same time, block printing on paper began to appear, principally originating from German monasteries, but occurring also in Venice and Flanders. Paper prints were not "decorative" artifacts of design or ornament, as textile prints still were. They were religious pictures: Bible scenes or portraits of the saints; early attempts at disseminating popular piety; holy pictures to hang in the kitchen. Gradually, as is evident in the wood block of St Sebastian, space was allowed under the illustration for the carving of a textual message. Eventually the paper prints from these blocks of picture plus text were made up into books. No surviving block book can be dated earlier than Gutenberg's first books printed from movable metal type. It seems that until the 16th-century block books and typographically printed books were produced and circulated alongside each other, the block books being distributed at a more popular level.

Some 1300 years after the introduction of the hand stamping of paper for visual design or verbal communication, a fourfold revolution elevated the craft of printing to a commanding height. *The method of production of the fifteenth-century European block book was basically not much different from that which had been carried out in the Orient for some hundreds of years*, says James Moran. *Typographic printing, on the other hand, with its special characteristics, can be truly said to be a European invention.* Principal among the innovators was Johann Gutenberg, a goldsmith of Mainz and Strasburg. In 1438, when he was about 40, he was working on the printing of books. By 1456 he had composed and printed a bound and illuminated 42-line Bible. His first major venture is still an aesthetic wonder, and established a technological process which was to last virtually unchanged for four-and-a-half centuries.

The four printing processes over which Gutenberg achieved simultaneous mastery concerned the manufacture of practical movable metal type, a new composition of ink, a technical understanding of paper, and the adoption of the printing press, an accurate mechanical device for transferring ink swiftly and cleanly to paper. Gutenberg started with the advantage of the then concise 23 letter European alphabet. But he found that, with capitals, punctuation marks, abbreviations, and multiple symbols such as *fi* he needed some 150 characters. Using his experience of the Mainz mint, he evolved a process by which a letter, engraved in relief on a steel punch, was struck into a slab of softer metal such as lead to make a matrix. The matrix was placed at the bottom of a mold extending upwards for about an inch. Molten metal – probably a tin-based alloy containing bismuth – was poured into the mold. When cold, the type was removed and the shanks were filed down to a uniform height, then the type was stored in cases. For composing, the letters were picked out individually and set in a stick to make a line. The lines were locked into a chase for printing the number of pages which the frame contained.

Before Gutenberg, printing ink had been water-based. This composition did not spread easily, particularly over the surface of metal type, and it tended to accumulate in droplets. Gutenberg used an oil-bound ink made by grinding boiled linseed oil with carbon. Flemish artists, who had previously painted in tempera – pigments ground with egg yolk and water – had only recently discovered the use of oil as a color medium. In 1434 Jan van Eyck painted in oil-bound pigments his masterpiece *Giovanni Arnolfini and his Bride*, depicting the betrothal ceremony between a merchant of Lucca and his delicate but heavily pregnant lady.

Gutenberg had to deal with paper which had been manufactured and starched, yielding a hard surface capable of being worked on with a quill pen without excessive absorption of the ink. In transferring the text from inked type to paper, he had to devise an entirely new alternative to the slow hand rubbing on the back of the paper which had been the previous method of printing. He quickly learned to dampen the paper so that it did not slide and took adequate absorption of ink. But there was no available tool which would provide the short, sharp, and fast contact between moist paper and oil-smeared metal which was necessary for the comparative mass production of the printing process.

Developing the printing press, Gutenberg possibly took his idea from the wine press, but his machine bore little resemblance to that massive model. He produced a dainty screw press worked by one pull on a bar handle which squeezed type and paper evenly, briefly, and with the calculated resilience needed to produce a clean, unslurred page printed in series at an acceptable speed.

Trial and error must have consumed much time and many spoiled sheets. Probably Gutenberg did not begin his Bible until 1454, and certainly it was not his first work. But the Bible that emerged was without blemish: 300 copies of a 1282-page book, some printed on paper, some on vellum. The effect of printing on scholarship, on science, and on technology may be deservedly noted in the words of Isaac Asimov: *The base of scholarship broadened and the educated community grew in numbers. Furthermore, the views and discoveries of scholars could be made known quickly to other scholars. Scholars began to act as a team, instead of as isolated individuals. The realm of the unknown could more and more be assaulted by concerted blows. Scientists were no longer fists, but arms moving a battering-ram.*

The first book printed with technical or scientific illustrations was Roberto Valturio's *De re militari*, a treatise on military technology from the point of view of the engineer and his commander, published in Verona in 1472. (It had been completed in manuscript in 1455, a year after the issue of the Gutenberg Bible.) The illustrations were done by the engraver and medallist Matteo dei Pasti. The book was written at the request of a sophisticated general, Sigismondo Malatesta, Lord of Rimini, who, as the illustration of the questionably seaworthy "monitor" suggests, had a lively interest in the assault and defense of coastal strongholds, having himself designed the fortifications of Rhodes and of Ragusa (modern Dubrovnik).

Valturio concerned himself with archaic as well as contemporary techniques in a sort of compendium of military engineering. His book was in such demand that four editions were called for in the next quarter-century; Leonardo da Vinci possessed a copy. At the time of the publication of *De re militari* Leonardo, only 20 years old, was an extremely accomplished painter. Later he was taken up by Lorenzo the Magnificent, the de' Medici magnate of Florence, for whom he worked indefatigably for four years, almost exclusively on engineering projects. For the rest of his life he divided his time between art and science, in the latter sphere most notably as chief engineer to Cesare Borgia, the brilliant and ruthless model for Machiavelli's *The Prince*.

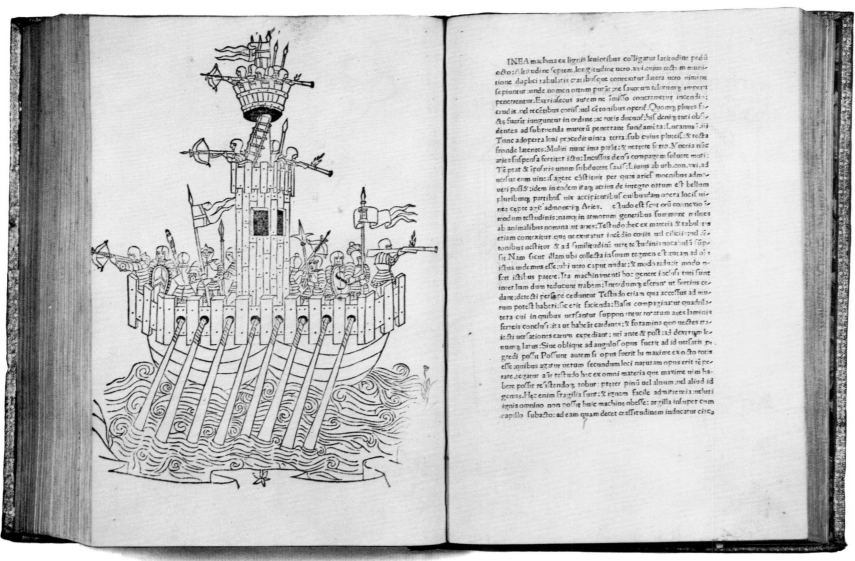

The design for a helical-screw helicopter from the notebooks of Leonardo was made between the years 1486 and 1490, when Leonardo was working as a military and civil engineer for Duke Ludovico of Milan. The helical design for the rotor is aeronautically inefficient. The power supply was intended to be from spring drive or clockwork. The instructions (in mirror-writing, which Leonardo seems to have practiced naturally) say that if the rotor is turned at great speed *this helix can screw into the air and rise high . . . A small model of this can be made from cardboard.* It is strongly believed that Leonardo made – and flew – a working model of this concept; if so, this was the world's first powered aircraft, carrying its own power unit and becoming airborne. The notebooks holding this design and many others were lost for nearly 300 years and not published until 1797, the year after Sir George Cayley had produced his own model helicopter. A century later, Leonardo's design was shown to Igor Sikorsky during his childhood in Russia. It so impressed him that, even though he failed for many years, Sikorsky was obsessed with the desire to design successful helicopters.

Leonardo's design for an ornithopter – a bird-like wing-flapping device into which a man could buckle himself and then cast off and fly – was an impossibility that was to be pursued into the 20th century. Leonardo planned that the man should lie face downwards inside the hoops, his hands operating the oar handles of the two angled spars of a broad and deep bird-wing system. Wire loops attached to the flyer's legs worked a tail unit. In other versions Leonardo depended on the pilot's knees flexing up and down to assist the beating of the wings, while the tail elevator/rudder was controlled by a head harness. Leonardo thought he understood how birds fly, saying that *the wings row downwards and backwards like swimming in water.* But bird wings drive downwards and forwards and the primary feathers at the wing tips actively screw into the forward air. Man has not the air frame or the muscles to fly like a bird. A bird has light, hollow bones, often internally strutted, a rigid spine, a deep keel of a breastbone allowing the adhesion of powerful flight muscles, which are fueled and cooled by auxiliary lungs, and its center of gravity places it in a far more efficient flying attitude.

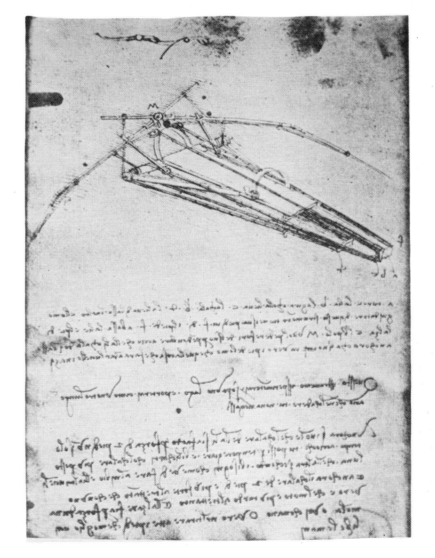

Euclid wrote his *Elements* of geometry and numbers in Alexandria in about 300 BC. The original Greek text was lost, but there was an Arabic translation. In the early 12th century Adelard of Bath translated the Arabic into Latin, and hand-written copies became prized by students. In 1482 Erhard Ratdolt produced the first printed edition at his press in Venice. It is a beautiful book, with elaborate borders and hundreds of woodcut initials, but it is outstanding as the first production of a long mathematical book illustrated by diagrams. The figures are printed from metal lines and laid out in the wide margins to enhance the beauty of the book.

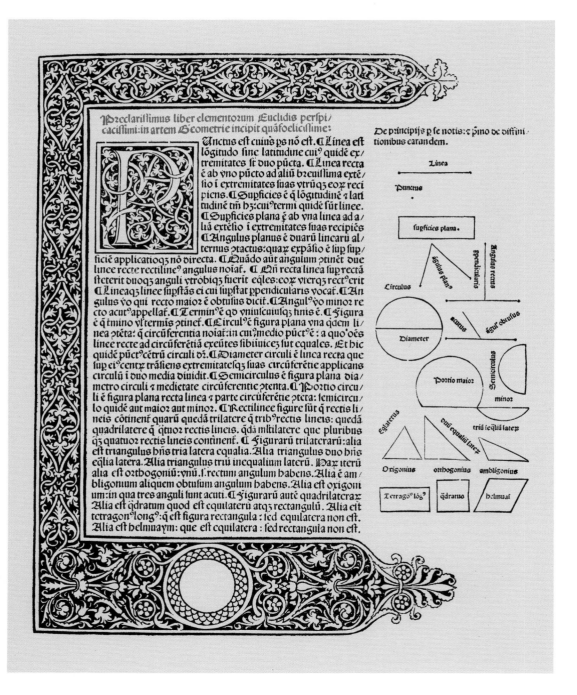

Euclid, preserved and re-presented to the world by Islam, was to be given 1000 further editions in the next four centuries. Eight years after the first publication, Aldo Manuzio founded a press, also in Venice, which was to influence the availability of classical literature on a more extensive scale. He was a classical scholar, tutor to the princes of Carpi, near Modena in Italy, who persuaded his patrons to set him up as a printer. Under his latinized name of Aldus Manutius he launched his press in 1490, and published the entire works of Aristotle, the first major Greek text ever to be printed. But he had always cherished the intention to publish a series of classical texts in small format – *portable or pocket books*, as he called them. Until Aldus the only small books had been prayer books. The Aldine *Virgil*, published in 1501, has a page area smaller than the modern standard paperback. It was only the first text of many. In what Harry Carter has called *a revolution in publishing . . . the house of Aldus published the first series of books uniform as to format, the classics the people ought to have read, reasonably priced ; and it kept the books in print.*

In the first century of printing, the ancient works of Ptolemy, Bede, Galen, and Vitruvius jostled with contemporary publications by Columbus, Machiavelli, and Copernicus as well as the controversial theologians of the Reformation. The metaphysical even made room for the profoundly physical. In 1556 a compendious handbook on mining, mineralogy, metallurgy, and foundry practice called *De re metallica* was published. It had been written by a Saxon physician and mineralogist named Georg Bauer who, like Aldo, latinized his name to Georgius Agricola. The book's woodcuts by Hans Deutsch include the first picture of a railway, and an accurate representation of a horse whim for raising

large loads. Precautions against a runaway hoist include a hook to stay the chain and a brake drum on the driving shaft against which a beam can be brought to bear through the weight and strength of the man on the lowest level. Agricola's clear exposition of mining begins the age of literate technology. Francis Bacon, who was an infant prodigy when the book was gaining influence, declared *I take all knowledge as my province.* But his was the intellectual wing of experimental science. At the humbler horn of the crescent there was the alert artisan described by Professor J. D. Bernal: *Printed books made it first possible and then necessary for craftsmen to be literate.*

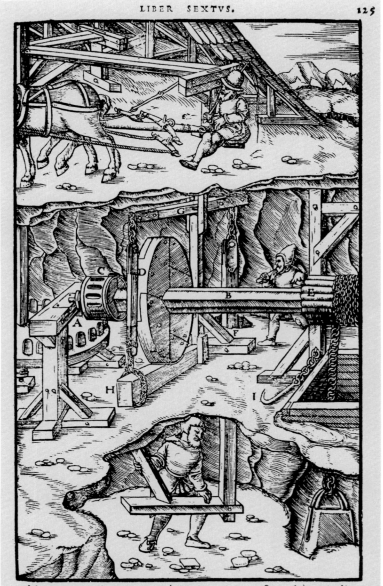

THE GALLEON

For many centuries Northern Europe had one tradition of navigation and shipbuilding, while Southern Europe had another quite separate one. From the middle of the 13th century, however, the two traditions came into increasingly close contact, largely because of the maritime route developed between the Mediterranean and the North Sea around the Iberian Peninsula. The result of this was a new type of vessel, the full-rigged ship, which in its various and successive versions was called the caravel, the carrack, the galleon, and the fluyt.

Genoese galleys had carried firearms in 1338; and Venetian boats, bombards (cannon) in 1380. The development of the full-rigged ship in the 15th and 16th centuries permitted more extensive use of artillery on board. At first, guns stood on the deck of the ship's castles. Later on, as guns became heavier, the larger ones were carried on the upper deck, where they fired over or through the bulwarks. At the beginning of the 16th century it was found that if ports were cut in the actual hull of the ship, then cannon could be mounted on the main deck. This discovery made it possible to mount more and bigger pieces without imperiling the stability of the vessel. Thus the ocean-going sailing ship became a floating fortress carrying hundreds of guns. This kind of ship was the technical contrivance that made possible the European overseas expansion.

Overleaf: Christopher Columbus' caravel *Nina*, from *Oceanica Classis*, printed by Johann Bergmann de Olpe, Basle, 1493

Powerful expansionist forces were at work in Europe as early as the 13th and 14th centuries, but they were not supported by adequate technology. The Italians had used wind energy and gunpowder, but only in a secondary role. Essentially they continued to rely on human energy for movement and fighting. But a crew could barely master the ocean with its oars. Any enemy with superior numbers was bound to prevail over them in hand-to-hand fighting. The gun-carrying, full-rigged ship allowed a relatively small crew to command unparalleled quantities of inanimate energy, both for movement and destruction.

In 1513 Albuquerque proudly wrote to his King that, *at the rumour of our coming the* [native] *ships all vanished and even the birds ceased to skim over the water*. Within 15 years after their first arrival in Indian waters, the Portuguese had completely destroyed the naval power of the Arabs and their King could justifiably style himself *Lord of the Conquest, Navigation and Commerce of Ethiopia, Arabia, Persia and India*. Before the non-Europeans had absorbed the shock of the first contact with the Atlantic vessels, the advance of European business and technology had produced more efficient and more numerous vessels. The caravels and the carracks were followed by the galleons. The Portuguese fleets were followed by the vastly more formidable fleets of the Dutch and the English.

While the Europeans had a relative advantage on the seas, they remained highly vulnerable on land. Their artillery was slow and difficult to move, and its rate of fire was too slow to cope effectively with large numbers of human attackers – a serious drawback, especially overseas where the Europeans were few and their opponents were many.

Until the 18th century, European possessions around the world consisted mostly of naval bases and coastal strongholds. Though it is not customary when drawing maps to paint the seas with the color of the predominant nations, only such maps would give a correct picture of the nature and extent of European dominance and the role of the Atlantic nations as world powers during the first centuries of the modern age.

On August 3, 1492 at sunrise Christopher Columbus slipped out of Palos harbor, near Cadiz, commanding two neat caravels from his flagship, the *Santa María*, a carrack whose sailing qualities he distrusted. On October 12 he made a landfall in the Bahamas, which he declared to be the East Indies, and on October 27 he disembarked in Cuba, which he said was Japan. Later he decided that the islands lay beyond Japan, off the coast of China, and he reported this to his sponsors, Ferdinand and Isabella, the rulers of Spain, when he returned six months later, without the *Santa María* and without ever having touched the American mainland. But in 1498 he did sight Venezuela, which was visited the following year by Amerigo Vespucci. The belief of Columbus that the "New World" was a part of the Asian mainland was incorporated into the first map printed after his voyages, engraved by Francesco Roselli and printed by Giovanni Contarini in 1506, the year Columbus died.

Columbus failed to have his name appropriated to America, but only to the historical period known as pre-Columbian America. He had "discovered" the New World only in the sense that no one in southern Europe knew that the Norwegian Vikings had founded settlements in the northern Continent by 1000 AD, and they had been long preceded by Mongolians who had crossed the land bridge over the Bering Strait in the final Ice Age. Columbus made false assumptions about the vicinity of the New World to Asia: but his seamanship and vision drove him and the other great Renaissance explorers, responding to the need for bullion and the incitement of trade and exploiting the seaworthiness of the new ships, to push into uncharted seas. The goal had always been China because Marco Polo had invested that country with the glamor of treasure and trade two centuries earlier. Columbus continually carried a manuscript of Polo's *Travels* and scribbled notes in the margin. Seven years after the death of Columbus Balboa crossed the isthmus of Panama and raised the banner of Castile over the Pacific. Six years later Magellan, who already knew the East Indies from sailing the well-charted route round Africa, began the voyage that was to map the west coast of South America, confirm the separate identity of the American continent, and continue westwards to the East Indies, from which one surviving ship sailed westwards home.

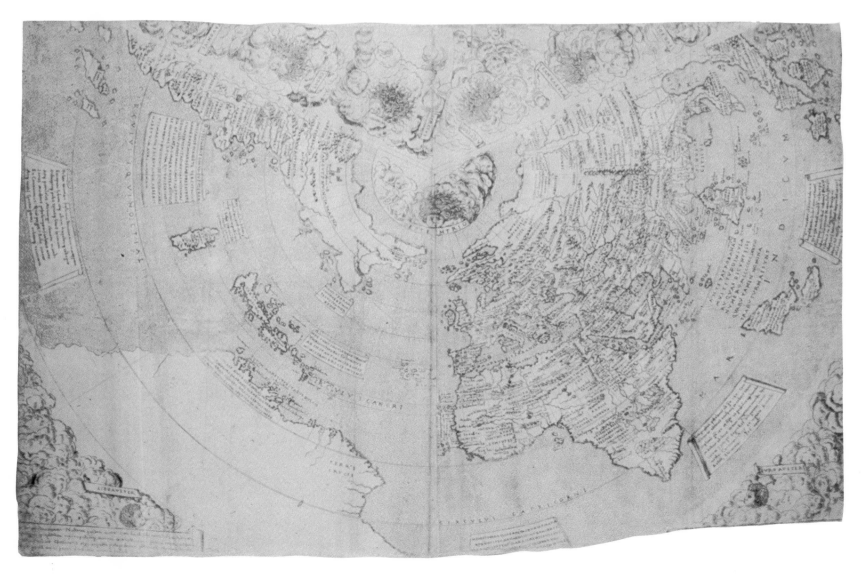

Maps are most natural and instinctive media for man's communication. No refined artistry was necessary for *Homo sapiens* to draw in the sand a directional diagram that was at least more durable and available for reference than a series of gestures. But a map of the world demanded greater grasp and more abstract projection. Paradoxically, such a map became more feasible only after maps of the heavens had been produced by astronomers trained in a discipline of accurate charting. No maps survive from the ancient Greek and Roman worlds, but the eight-tome *Geography* by Ptolemy of Alexandria, written about 150 AD, summarized the knowledge of the time, located thousands of geographical features, and laid down a system of cartography based on a spherical Earth. The tardy republication of this work, printed in 1475 with new maps based on an interpretation of Ptolemy's text, persuaded Columbus that he could reach China by a westward voyage. The projection of a globe on a flat map was a problem which became the life work of Gerhard Kremer, an instrument maker and graphic engraver from Flanders who latinized his name to Gerardus Mercator. He produced a map of the world in 1538 using a double-cordiform projection with each terrestrial pole at the centre of a separate heart-shaped form. He alternated his production between terrestrial and celestial globes and a succession of small-scale wallmaps based on accurate surveying: 15 sheets for Europe, eight sheets for the British Isles. For long-distance nautical navigation there was still no accurate map in existence. The portolano maps using compass bearings covered comparatively small areas or, in the case of the entire Mediterranean, were confined to a narrow range of latitude. To chart, say, a northwesterly course across the Atlantic encountered the problem of allowance for the convergence of the lines of longitude, because of the different distances between meridians at different latitudes. Mercator adapted a cylindrical projection which distorted the shape of areas of land and sea progressively as the latitude approached the poles. Manipulating this projection, to which his name was later given, he published in 1569 his famous map of the world which he specifically described as *designed for the use of navigators*: The virtue of his projection was that, even over a vast distance, the *rhumb-line* (the line followed by a ship sailing in a fixed direction) was a straight line on the Mercator map. Navigation became, as it had roughly been with small-scale portolano charts, a matter of straight-edged rulers.

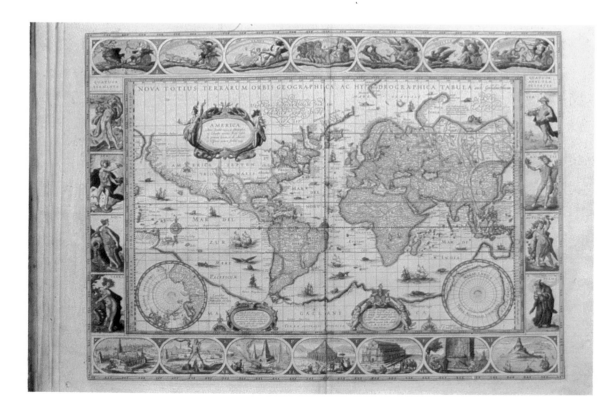

127

The Julian Calendar, established by Caesar in 45 BC, had been a serious attempt to forecast the seasons for agricultural and religious purposes by reference, not to the lunar cycle, but to the solar year, the time taken by the Earth to make one revolution round the Sun. But its compilers did not take the logical step of centering on the Sun that small part of the Universe we now call the solar system. Traditionally, from Pythagoras in 500 BC through Plato and Aristotle to Ptolemy, it had been postulated that the heavenly bodies moved around the Earth in perfect circles within their individual perfect spheres, (given an imperfect eccentric motion by Hipparchus in the 2nd century BC), making celestial music as they swung.

The Ptolemaic system was preserved by Islamic scholars and finally transported to the West, where it acquired immediate acceptance, when Ptolemy's *Almagest* was translated from Arabic into Latin in 1175, just 11 centuries after his birth. After an initial period of idolatry Ptolemy's system began to be recognized as astronomically unreliable. Long-term predictions of planetary movements were faulty, even when incorporating the corrections of Alfonso the Wise. Alfonso, king of Castile from 1252 to 1284, had made the historic criticism of Ptolemy: *If God had asked my advice during the Creation I should have recommended a simpler system.* A conference held in Rome in 1500 to consider calendar reform was attended by Nicolas Copernicus, a 27-year-old Polish astronomer, who began to reflect that the *simpler system* would obtain if the Sun was taken as the center of the then categorized universe, and the Earth was a planet of the Sun. Between 1512 and 1530 he developed his theory, at first purely as a mathematical convenience to forecast planetary positions. But Copernicus proved too shrewd a guesser. Like a cryptographer trying random combinations, he came near the truth without exposing the whole truth.

Copernicus recast the known Universe, setting the Sun in the center, with Mercury, Venus, Earth, Mars, Jupiter, Saturn and the "sphere" of the fixed stars revolving around it. The notion that the Earth actually *moved* was a theological shock of fundamental impact. Martin Luther, 10 years the junior of Copernicus, opposed the theory as firmly as did the more orthodox disputants in the Vatican. In 1543, the year of his death, Copernicus' book, which had been circulated in manuscript among the intelligentsia of Europe, was published as *De Revolutionibus Orbium Celestium* (The Revolution of the Heavenly Bodies).

The doubly revolutionary book did not make converts quickly. As Galileo was to discover, its tenets were officially deemed heresy. The most brilliant of those who disagreed was Tycho Brahe, the amazingly accurate "naked-eye" astronomer (he died eight years before telescopes were introduced). Tycho's precise eye and his mastery of instruments, bolstered by royal patronage which bought him a magnificent observatory, enabled him to prepare new astronomical tables determining the length of the year to within a second. Clearly, it was time for calendar reform. The Julian year was running 10 days late.

In 1582 Pope Gregory XIII sanctioned the Gregorian correction to the Julian calendar, shifting the date forward 10 days and abolishing leap years in those centennial years not divisible by 400. The four principal Catholic countries of Europe complied immediately, in October, 1582, and the remaining Catholic countries agreed in 1583. Queen Elizabeth I of England ignored the Pope, who was backing Mary Queen of Scots as her successor at the time. Britain and America did not change until September, 1752. Protestant Germany, the Netherlands, and Denmark complied in 1700.

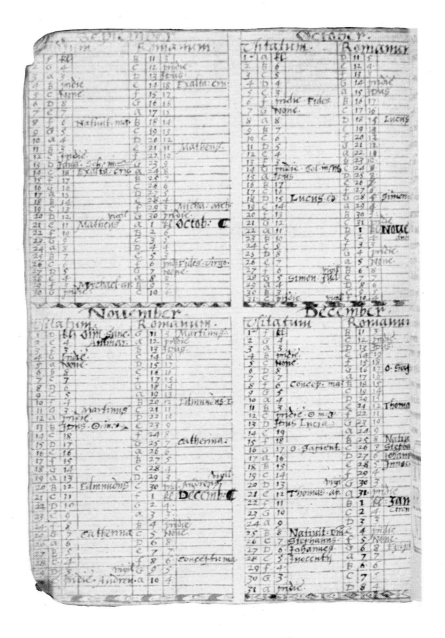

By the end of the 15th century the design of the sailing ship had developed into the form of the full-rigged ship from which it did not greatly vary for 300 years. Over the previous two centuries the styles of Mediterranean and northern ocean craft had merged. Adoption of the triangular fore-and-aft lateen sail allowed vessels to beat into the wind. Adaptation of the high sloping stern line to take a vertical stern post and a central stern-post rudder enabled a helmsman to control his course in a ship responding far more readily to it than to the old steering-oar on the starboard quarter of the stern. Better steering led to finer seamanship, more effective tacking and more accurate navigation over the longer distances which trading opportunities suggested and the mariner's compass permitted. The oceans were open, and more capacious, more effective ships were developed to sail them. Capacity, more often envisaged for cargo, had in a proportion of any fleet to be devoted to armament. The "Great Harry", *Henry Grâce à Dieu*, launched at Woolwich in 1514 as a Royal Naval showpiece and, like many a deterrent, never tested to extremity in action, was a carrack of 800 tons built with eight decks and requiring a crew of 900 men to sail her and work the 186 guns mounted aboard her, some of which weighed over two tons. The Pepysian Library record shows her after rebuilding and refitting in 1539. The second mizzenmast, called the *bonaventure*, on the stern castle was an innovation of the 16th century.

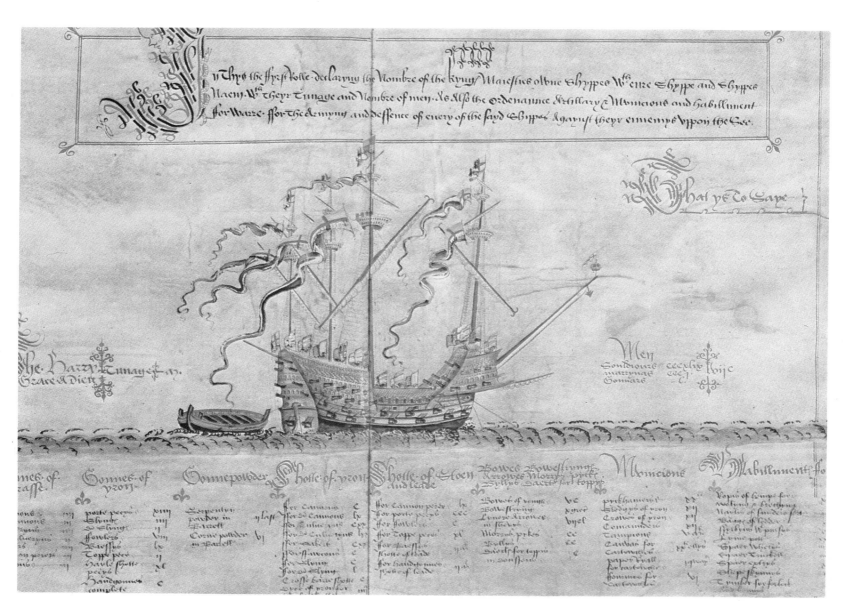

The first *Ark Royal*, the flagship of the Lord High Admiral Baron Howard of Effingham during the engagement with the Spanish Armada in 1588, was a galleon displacing 800 tons and requiring a crew of over 500 men. The galleon was a comparatively slim craft, its length four times its beam, highly manoeuvrable though packed with artillery, having the castles fore and aft much lower than in its clumsy predecessors. Its architecture was governed by its function as a specialized warship, its tactical role flexible but most effective in attack. Skilfully sailed by the English and the Dutch, the English being particularly accurate gunners, the galleon changed sea warfare from the old cycle of confrontation, grappling, boarding and pitched infantry fighting to the crushing impact of heavy artillery fluidly used in the running broadside and speedily repeated from another quarter.

The Dutch used galleons to challenge and subdue Portuguese naval power in the East Indies and intervening areas. Their particular contribution to the design of merchant ships, deriving from conditions in European waters, was the *fluyt*, with a characteristic pregnant line and narrow stern. It was originally designed for purposes of tax-avoidance: the states controlling the lucrative Baltic trade imposed a tax based on the deck area of the carriers. Consequently the fluyt had a hull curving inwards to the top deck, on which the fiscal assessment was made. The fluyt was first built in 1595. The design proved to have produced very fast ships of high capacity, and it was widely adopted.

The voyages of Sir Francis Drake made him an explorer almost by accident, but an adventurer fitting and decorating his age like a historic symbol. Drake, having earned a formidable reputation and a considerable fortune as a buccaneer licensed to harass the Spanish and maraud their treasure fleets, was given an extraordinary assignment in 1577. Queen Elizabeth's secretary of state, Sir Francis Walsingham, commissioned him to sail through the Straits of Magellan and pass north to attack Spanish ports and shipping along the coast of Peru. He had a highly remunerative series of encounters on a voyage extending as far north as modern Vancouver, where he effectively disproved the fancied existence of a North West Passage to Asia, between America and the North Pole. He refitted at San Francisco, where he claimed California for the English Crown as *Nova Albion* in 1579.
He sailed to the Moluccas and across the Indian Ocean, completing his circumnavigation of the globe with his arrival at Plymouth in the *Golden Hind* on April 4, 1581. His career as an individual marauder of the Spanish homeland and empire continued until he died of dysentery in the West Indies in 1596. Apart from his personal achievement in the team that battled with the Armada, Drake did much to establish the future supremacy and traditional morale of the Royal Navy. His resource, professional skill, idealism and intermittent brutality stamped him as a characteristic Elizabethan hero who *put a Girdle about the World*.

Among the navigational instruments developed during the 16th century was the nocturnal, enabling mariners to tell the time by the stars. The specimen by Humphrey Cole was made while Drake was on his voyage round the world. It converts an observation of the comparative positions of the Pole Star and Kochab, the bright star at the head of the Little Bear, into an indication of the time of night. The serrated half-hour graduations on the revolving inner disk permitted the time to be told by touch. The nocturnal reads 10 pm on May 27. Many nocturnals were backed with a tide calculator and lunar dial indicating periods of moonlight. The stars were to continue to rule navigation through the centuries, but far more detailed study of them was soon to be afforded by the introduction of the telescope.

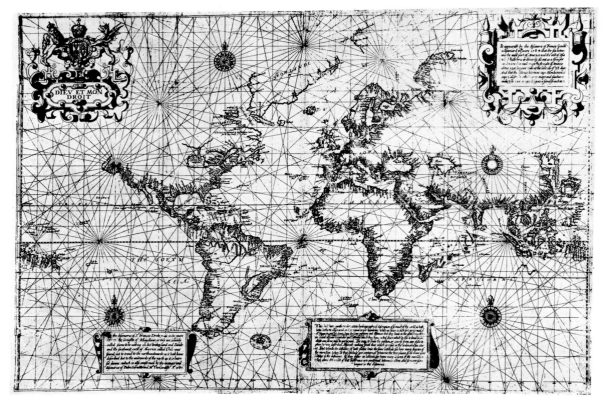

THE SCIENTIFIC REVOLUTION

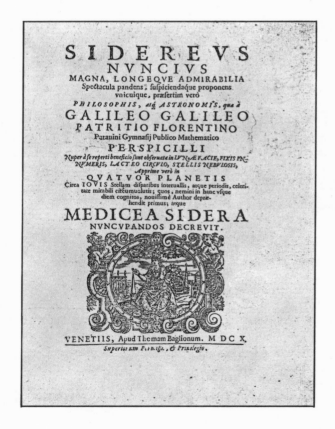

Technology had traditionally been the realm of craftsmen working by rough rules of trial-and-error, *their knowledge lying in their hands confusedly*, as Samuel Pepys so neatly put it. Official, proper science did not concern itself with technology. Both in the classical West and in classical China the man of *science* was the man of knowledge – the philosopher. The clear division between science and technology was further emphasized by the vast social difference between the upper-class philosopher and the lower-class craftsman. One could work with the mind while the other had to toil with his hands.

Plato's scorn for the tiny group of philosophers who ventured to devise mechanical aids for the solution of problems in geometry mirrored the typical prejudice of the landed gentry and the intellectuals of antiquity against mechanical gadgetry. The Alexandrian school of engineers devised a hodometer: a carriage that measured distance traversed and transmitted these measurements to the coachman. But Roman land surveyors never used this ingenious tool. The Emperor Commodus left a hodometer among his possessions, but his successor disparagingly discarded it as just another toy. Fifteen centuries later, in a similar vein the compilers of *Sau-ku ch'uan-shu ti yao* wrote that the mechanical and scientific instruments brought to China by Westerners were *simply intricate oddities designed for the pleasure of the senses*.

Overleaf: Title page to Galileo's *Starry Messenger*, printed in Venice, 1610

They fulfil no basic needs. Manual labor and material production were for the lower orders, and beneath the dignity of the leisured ruling class. The role of this class was to rule people, not manage things, so their interest in mechanical contraptions would be regarded by their peers as a childish fascination with toys.

This lofty disregard for mechanical devices began to wane in medieval Europe. The success of the water mill after the 8th century AD not only was the first breach in the energy bottleneck, but it also marked the first interest in mechanical devices by the wealthy and the powerful, who had until then neglected technological progress. What had started as a practical development soon gained intellectual credence.

By the 13th century there were some philosophers in Europe, such as Grosseteste and Roger Bacon in Oxford, who did not consider it beneath their dignity to apply their minds to technological matters and began to think about practical experimentation. But this new outlook met with stubborn and widespread resistance in the world of scholarship; as the monk Abelard wrote, those who *concerned themselves with action* might accomplish useful things, but were in principle no different from the beasts *which are sound in practice but ignorant of nature and cause.* The theoretical sciences had dealt for 16 centuries only with matters unchangeable and eternal – first causes and forms of being – which, precisely because they were unchangeable, could not be harnessed to action. The task of the philosopher and scientist was to educate the elite how properly to govern and to order society, not how to play with mechanical contraptions.

We are accustomed to think of the Renaissance as a period of cultural revival which fostered technological inventiveness and change. Certainly, clearer, more rational attitudes did create a background more favorable to systematic investigations. But the classical values which molded the Italian Renaissance were not favorable to applied technology. On the contrary, they emphasized the old distinction between science and technology: philosophical speculation was considered an attribute of gentility, while concern with practical matters was seen as a mark of vulgarity. Physicians strove to rank themselves as philosophers and to dissociate themselves from the surgeons, who were looked upon as technicians and therefore as low-class artisans. Architects worked tirelessly to dissociate themselves from engineers.

However, the legacy of Grosseteste, Bacon, and Dondi was not lost. The development of an urban, mercantile economy ultimately favored a utilitarian view of science and fostered a rapprochement between science and technology.

By the end of the 16th century, a reaction to the classical view was in the air. It gained momentum in the following century when it reached such an intensity and a diffusion that it came to be known in historical writing as the Scientific Revolution.

Galileo wrote that he cared *more about small facts which could be tested than about the great questions which could be neither proved nor solved*, and to Bacon *the true end of knowledge is not the pleasure of the mind* but *a line and race of inventions that may in some degree subdue and overcome the necessities and miseries of humanity*. The cultural tradition which had reigned supreme from Aristotle to Thomas Aquinas was being challenged at its very foundations. Nor was the struggle between the two schools as calm and detached as the passages above might lead one to believe. More often than not its tones were vitriolic, even violent, for two mutually incompatible cultures were fighting for survival. If we tend today to share Bacon's view and not Abelard's, it is because in their fierce cultural struggle in the 17th century the "moderns" won over the "ancients".

The reaction against traditional values and the desire to impose the experimental method led the moderns to a radical reappraisal of the work of craftsmen. Bacon repeatedly emphasized the need for collaboration between scientists and artisans. Galileo recognized his debt to the craftsmen of the Arsenal in Venice. The Royal Society of London directed some of its members to compile a history of artisan trades.

While all this was happening in the field of science, convergent developments were taking place in the field of technology. Protestantism and the printing press combined to encourage the spread of literacy among craftsmen. At the same time developments in oceanic navigation, artillery, horology, and the manufacture of precision instruments gave rise to a class of superior artisans capable of conversing with contemporary scientists.

Until the middle of the 19th century, the contribution of science to technology was to remain occasional and of little note. But the cultural and social revolution of the 16th and 17th centuries created the precondition for that union of science with technology which was to become the basis of modern industrial development.

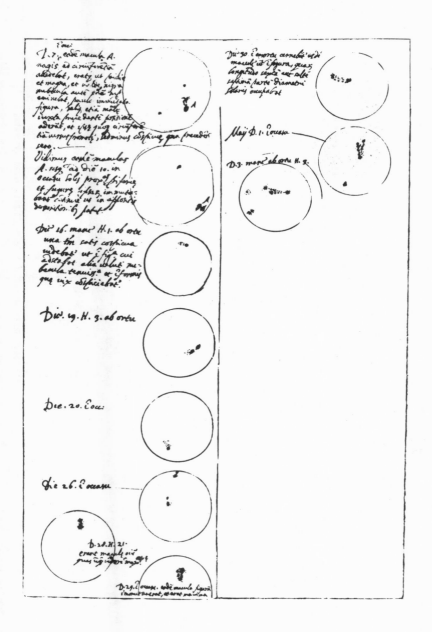

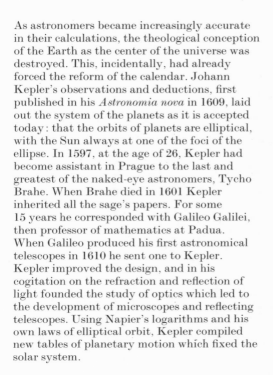

As astronomers became increasingly accurate in their calculations, the theological conception of the Earth as the center of the universe was destroyed. This, incidentally, had already forced the reform of the calendar. Johann Kepler's observations and deductions, first published in his *Astronomia nova* in 1609, laid out the system of the planets as it is accepted today: that the orbits of planets are elliptical, with the Sun always at one of the foci of the ellipse. In 1597, at the age of 26, Kepler had become assistant in Prague to the last and greatest of the naked-eye astronomers, Tycho Brahe. When Brahe died in 1601 Kepler inherited all the sage's papers. For some 15 years he corresponded with Galileo Galilei, then professor of mathematics at Padua. When Galileo produced his first astronomical telescopes in 1610 he sent one to Kepler. Kepler improved the design, and in his cogitation on the refraction and reflection of light founded the study of optics which led to the development of microscopes and reflecting telescopes. Using Napier's logarithms and his own laws of elliptical orbit, Kepler compiled new tables of planetary motion which fixed the solar system.

Galileo, who was born as Michelangelo died in 1564, spent 30 years in experiments in the field of physics and mechanics which toppled traditional Aristotelian views. As the pioneer of the telescope, he became the best lensmaker in Europe. He first used his numerous and beautifully cased telescopes to trace the course of the spots he observed on the Sun. Through his still existing sketches he showed that the Sun rotated on its axis every 27 days. He recorded the series of discoveries, which in effect corroborated the Copernican notion that the Earth with the planets moved around the Sun, in issues of his periodical *Sidereus Nuncius* (Starry Messenger) from 1610. But he delayed publication of his fully argued revolutionary theory until 1632 when he issued, in vernacular Italian, his justification of Copernicus against Ptolemy, the *Dialogue on the Two Chief World Systems*. A year later, when he was 69, he was forced to recant the heresy. He was old, and blind through damage done to his eyes in his reckless observations of the Sun, when the poet John Milton visited him in 1638. Milton, who would later become blind himself, and a prisoner of conscience, honored Galileo in his *Areopagitica*, and later, in *Paradise Lost*, acknowledged Galileo's perception of the luminous Milky Way as a multitude of stars invisible to the naked eye with his reference to

The Galaxy, that milky way
Which nightly as a circling zone thou seest
Powdered with stars.

John Napier, Laird of Merchiston near Edinburgh, was, like his contemporary Johann Kepler, passionately interested in the twin subjects of astronomy and religion. These studies have run in parallel courses as long as men have attempted to probe the mystery of Creation through investigating the reality of the universe. When Kepler first perceived through Galileo's telescope heavenly bodies he had not previously been aware of, he cried: *O God, I am thinking Thy thoughts after Thee.* Even in Napier's time, astronomy demanded endless calculation. The Scot set out to devise a simpler method of computation. He saw that all numbers could be considered as exponential expressions – as the *power* (the square, the cube, or a fractional exponent between or above the square and the cube) of a base number. By adding the exponents he could multiply, and by subtracting he could divide. After 20 years' work Napier published his tables of logarithms – or, *proportionate numbers.* The system was, with Napier's consent, almost immediately simplified between 1614 and 1624 by Henry Briggs, professor of astronomy at Oxford, who recast the logarithms as exponentials of the base of 10. The system of calculation by logarithms was speedily taken up by mathematicians, and Kepler used the new logarithms to finish his tables of planetary motion.

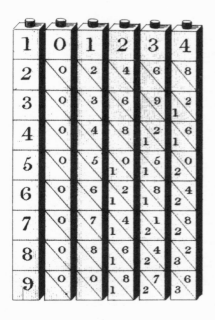

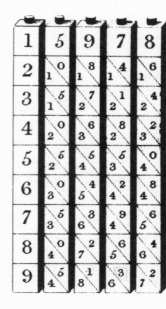

Napier then devised a system for multiplication with the box of cubes called *Napier's bones.* In the right-hand stack illustrated the first vertical column lays out the multipliers and the other columns the multiples. The top line is the multiplicand, here 5978. A total is found by adding the figures in the same diagonal strip. *Napier's bones* were quickly supplanted by the slide rule, using linear intervals based on logarithmic values.

The first design, consisting of two rules which had to be held together, was introduced in 1622 by William Oughtred, rector of Aldbury, Surrey. A circular model was propounded by Richard Delamain in 1630. But the master model for all modern slide rules, with the slide moving through a fixed stock, was introduced by Robert Bissaker in 1645.

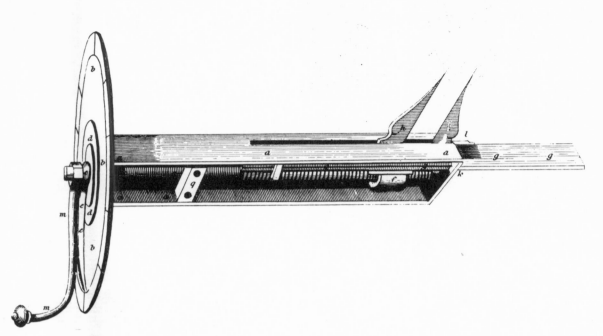

William Gascoigne, a youthful prodigy in optics and astronomy, lived and died in Yorkshire, being killed at the age of 22 at the battle of Marston Moor in 1644, where he fought on the royalist side. His micrometer, which he had introduced five years earlier, used the new principle of fine screw adjustment to focus the eyeglass of a telescope and to measure the angles subtended by parts of the image. With this instrument he measured the diameter of the Moon and of the planets, and Hooke later used modifications of Gascoigne's micrometer in microscopes and telescopes. Accuracy of measurement was thus first used exploratively, being applied to finding out more surely about the world and about more distant Creation. This was its main function during the scientific revolution. Two centuries later, during the Industrial Revolution, its application was to be directed more typically towards perfecting the manufacture of identical objects and the fabrication of interchangeable parts.

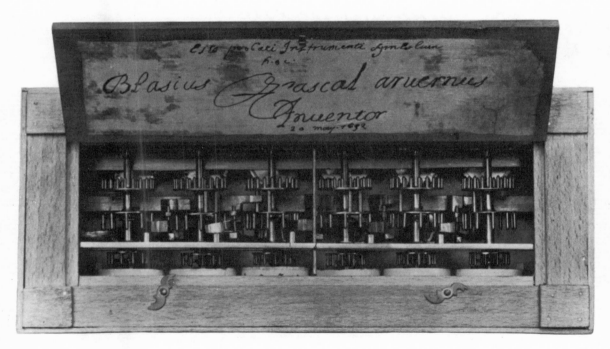

Blaise Pascal lived for only 39 years and abandoned mathematics and physics for the last eight of these. He was the brilliant son of an Auvergne family of lawyers and government functionaries. At the age of 16 he wrote what is still the classic treatment of conic sections. In 1642, at the age of 19, he devised an adding machine to ease his father's problems as a tax administrator. He freely distributed a number of these models, sending one to Queen Christina of Sweden, a young patron of learning. Working with Pierre de Fermat, Pascal used the data of a gambling acquaintance to produce his most durable mathematical work: the modern theory of probability which is still used by scientists to draw valid conclusions from a multiplicity of controlled, random observations. He also continued the work of Torricelli in determining the weight of air and thereby measuring altitude by barometric pressure.

Evangelista Torricelli went from Rome to Florence at the age of 34 to be Galileo's assistant during the old man's last days, and he succeeded his master as Professor of Mathematics there. The dying Galileo had suggested that he should investigate the weight of air. Next year, in 1643, he inverted a tube of mercury with its mouth in a bowl of mercury and produced the Torricellian vacuum at the top of the tube. The mercury did not totally spill because of the counterweight of the air pressing on the surface of the bowl. (The apparatus with the flaccid and inflated bubble gum is a variation of the experiment to show how a small amount of air inside a bladder expands as a vacuum is formed.) Torricelli observed that the height of the column of mercury was not constant every day but varied, presumably as the weight of the atmosphere varied. He had, in fact, fortuitously devised the first barometer.

The properties of the vacuum were popularized by Otto von Guericke of Magdeburg, whose theoretical understanding had been weak until he read of Torricelli's work after the latter's untimely death in 1647. Guericke served as a military engineer during the Thirty Years War and as a civil engineer afterwards during the reconstruction of the city of Magdeburg, where he became Mayor in 1646 at the age of 44. In 1650 Guericke built the first air pump and used it to extract air to make an efficient vacuum in a container. With true showmanship, he demonstrated that an air pump could form a vacuum in a cylinder, despite the counter-efforts of 50 men heaving on ropes fixed to the piston in the top of the cylinder. His most famous experiment, carried out as a spectacle for the Emperor Ferdinand III, was to put together two metal hemispheres (the Magdeburg Hemispheres), evacuate the air within, and have two teams of horses straining ineffectually at each hemisphere to separate the sphere. In addition, Guericke did valuable work in astronomical theory and experiments with static electricity.

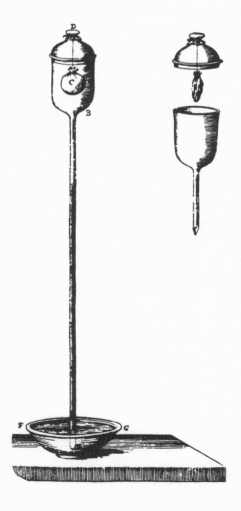

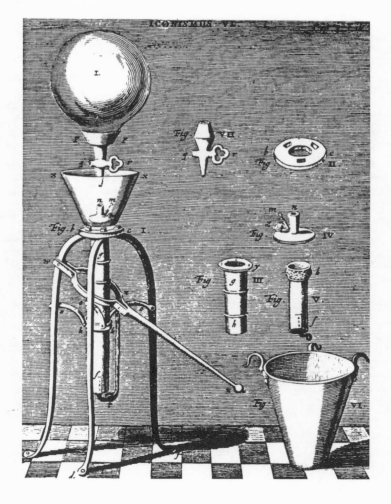

Robert Boyle, son of the Earl of Cork, was living in Florence when Galileo died there. He was only 15 years old, but had left Eton at 12 as an infant prodigy and had already read the works of Galileo during his Grand Tour of Europe. Back in England, Boyle virtually founded modern chemistry by defining the *element* as it is understood today – not merely as one of the four forces of air, fire, earth, or water. Boyle joined with Christopher Wren and other mathematicians to form a scientists' forum in London and Oxford which speedily became, with the restoration of Charles II, the Royal Society. In 1664 the King, their patron, *mightily laughed* at the Royal Society, Pepys reported, *for spending time only in weighing of air, and doing nothing else since they sat*. Boyle, having constructed an improved air pump with the aid of the dexterous Robert Hooke, had formulated "Boyle's law" on the relationship between the volume, density, and pressure of gases, which was to give a theoretical basis for the development of the steam engine. Meanwhile, the principle of the vacuum was being applied to the tentative science of aeronautics.

In 1670 the Jesuit Father Francesco de Lana de Terzi, sometime Professor of Mathematics at Ferrara, made a theoretical projection utilizing the air pump Guericke had demonstrated 20 years previously. De Lana published in 1670 the first volume of an intended encyclopedia of science. Two chapters were devoted to a description of a flying ship, suggesting that four evacuated spheres made of sheet copper, each 20 feet in diameter but only 1/250th of an inch in thickness, would be so buoyant in air that they would support five men in a boat that would sail through the sky. Philosophers (i.e., scientists) of the time were greatly impressed by this proposal until it was calculated that the paper-thin walls of the globes would collapse under atmospheric pressure, and could not be made thicker and stronger without becoming ineffectual through overweight. De Lana was secretly relieved, since he foresaw furious aerial warfare, by bombs and incendiaries as well as the airlift of invading forces, as practical consequences of his airship.

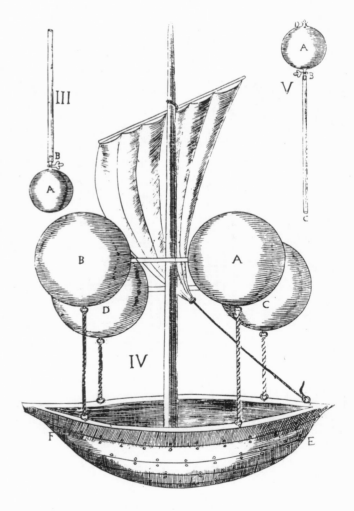

As a physicist and astronomer with the gift of being clever with his hands, Christiaan Huygens of The Hague gave 17th-century science an unparalleled leap forward in the efficiency of measurement and the resulting capacity to handle matter, energy, and light. His ability at grinding lenses led to his construction in 1656, at age 27, of telescopes with which he discovered unknown characteristics of Saturn and Mars. Two years later he devised a micrometer which permitted him (and the French astronomer Jean Picard, who used the innovation) to measure very small angles in an arc and thus, with the micrometer mounted on a telescope, to make accurate measurements of space. Picard's most influential use of the micrometer was to make a reliable calculation of the circumference and radius of the Earth. Huygens then set about achieving an accurate measurement of time. He used the pendulum, which Galileo had already established as having a regular beat. But the construction of a successful timing device needed both Huygens' mathematical theorizing – which showed that for absolute accuracy the pendulum's arc had to be slightly off the path of a perfect circle – and his mechanical ability. He was then able to devise a practical attachment at the fulcrum which achieved this arc. Using these skills, he constructed the first long-case clock in 1657, giving physicists new power to record time with accuracy.

Huygens used his lensmaking skills to construct better telescopes after the model which had already been introduced by Johannes Hevelius of Danzig. Hevelius, who was another competent lens grinder using his own new-style lathe, had established in the 1640s the best-equipped astronomical observatory in Europe. Specializing in very long telescopes of comparatively high power, he published in 1647, when he was 36, a detailed atlas of the Moon entitled *Selenographia*. He also produced two exhaustive works on comets and a reliable catalog of 1,500 stars. Eventually the long telescopes favored by Hevelius and Huygens were reaching lengths of up to 200 feet, with powerful rigs to support them. Edmund Halley, the first astronomer to make a record of the stars of the Southern Hemisphere, based on observations from St. Helena from 1676 to 1678, visited Hevelius on his return. Spurred by Hevelius' interest, Halley too began specializing in comets, which led him to predict the reappearance in 1758 and 1759 of the comet he had logged in 1682. When it finally reappeared in 1835 and 1910, it was called *Halley's Comet* (the same comet of 1066 which is depicted on the Bayeux Tapestry).

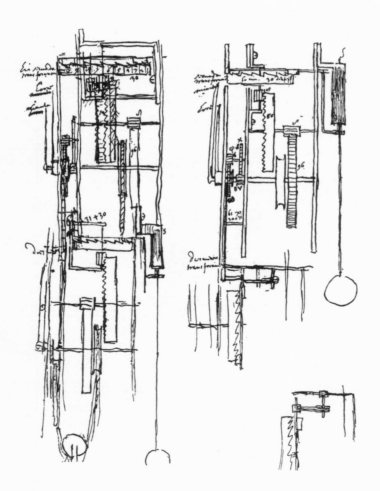

Huygens produced lenses only for himself and his friends, not on any commercial scale. Commercial lens grinding became the specialty of Ippolito Francini, who worked in the palace workshops of the Medici princes in Florence. Before his death in 1653 he was using a modified lathe to produce higher quality lenses for telescopes built on Galileo's model. A far more sophisticated lens-grinding lathe was constructed by Giuseppi Campani of Rome, who discarded the old practice of grinding and polishing a disc of glass which had previously been cast into a mold, and instead cut the lens directly from block glass, grinding and polishing as the work proceeded.

Isaac Newton was born in the year of Galileo's death. Twenty-four years later, the same year that Milton published *Paradise Lost*, the young Newton succeeded Kepler as professor of mathematics at Cambridge. Inspired by Kepler's work on optics and Robert Boyle's recent experiments on color, Newton gained prominence through use of a prism which refracted a white beam into the spectrum band. This prism extended his logical conclusions about the properties of light into realms of abstract mathematics.

Newton's worldwide renown as the great intellect of mathematical and scientific theory tends to obliterate his considerable skill as an artisan. Though weak and sickly all his life, his main activity as a boy was experimentation with homemade mechanical toys. Parallel with his early theoretical work on gravitation, prompted by the genuine physical circumstance of the falling apple, he was conducting practical experiments on optics. Recalled to Cambridge soon afterwards, he continued his work on light and color. His refraction of light with the prism had convinced him that *chromatic aberration*, the blurring and color confusion of images as seen through optical lenses, discouraged any further advance in astronomical observations using existing telescopes. In 1668, therefore, he produced a reflecting telescope, in which a small parabolic mirror concentrated the image before it was presented to the eyepiece.

Newton's first reflection telescope had a magnification of about 32, and was succeeded by an improved model in 1671. The theory of the reflection telescope had, in fact, been published by the Scottish mathematician James Gregory in 1661, but its construction by a London optician was a disastrous failure, and the first Gregorian telescope was finally presented to the Royal Society by Robert Hooke in 1674. Gregory had corresponded with Newton in 1672 and 1673 to compare the merits of their respective telescopes. But in 1675, when Gregory was professor of mathematics at Edinburgh, he was struck blind by a disease of the optic nerve while showing his pupils the satellites of Jupiter, and he died three days later. Ninety years after the production of Newton's first telescope, the London Huguenot optician John Dollond produced an efficient achromatic refraction telescope by combining different kinds of glass in the lens with varying refractive qualities. The innovation was highly significant to microscopy, in which the refraction principle had never been successfully replaced.

Microscopes were developed at the end of the 16th century as a by-product of the spectacle-making craft then localized in Holland. Galileo realized that the telescopes he had been producing from 1609 and 1610 could be used as microscopes when reversed, but he seems to have done little about it. The development of microscopy as a field of serious research dates from 1653 with the Bolognese physician Marcello Malpighi, who from his study of frogs made important discoveries on respiration and the circulation of the blood. But the most widely appreciated work in the field was done during the next 10 years by Robert Hooke. In 1665 he published his *Micrographia*, abounding with the most delicate illustrations (including a record of Hooke's own microscope), some of them probably the work of Christopher Wren.

Robert Hooke was a peculiar polymath with an unpleasant personality, a wide-ranging mind, but a certain lack of intellectual stamina as evidenced by the astonishing number of scientific truths which he formulated but never proved conclusively. It was said of him that he *divined before Newton the true doctrine of universal gravitation, but wanted the mathematical ability to demonstrate it.* This criticism applies in many other spheres where he claimed pioneering ideas which were nevertheless attributed to his contemporaries. These include the properties of air with regard both to respiration and combustion, the periodicity of comets, and the physical cause of earthquakes.

Left as an orphan with a legacy of £100 at the age of 13 in 1648, Hooke came to London from the Isle of Wight and apprenticed himself to the portrait painter Sir Peter Lely. He quickly switched to academics at Westminster School and Christ Church, Oxford, where he was a servitor, or menial scholar. Robert Boyle took him up and used his skill to construct the air pump. By the age of 20 Hooke was accepted into the fellowship of Boyle, Wren, John Wilkins (an early aeronautical theorist, to whom Hooke confided *thirty several ways of flying*) and others who formed the Invisible College which soon became the Royal Society. Hooke was not only a theorist. As early as 1658 he devised the balance spring which gives watches a regular beat, and this hairspring principle was quickly taken up by Thomas Tompion. (Hooke gave his only intimate friend, Sir Christopher Wren, a grand Tompion watch which cost him £8.) *Hooke's Law* in physics refers to the elasticity of springs. Hooke, who lived penuriously, was given a salary as permanent curator of experiments in the Royal Society, of which he became a fellow in 1663. Samuel Pepys, discussing the savants of his time, placed Hooke above even Boyle, saying: *Mr Hooke, who is the most ... promises the least, of any man in the world.*

Although Pepys always enjoyed the *mighty discourse* he had with Hooke, he was at times overwhelmed by some of Hooke's claims (one true claim, which Pepys distrustfully termed *a little too much refined*, was that Hooke could ascertain the musical note registered by the buzzing of a fly's wings and the number of vibrations the wing was making every second). After the publication of *Micrographia* in January 1665 Hooke was appointed professor of geometry and lecturer in astronomy at Greshams College, London. After the Great Fire of London in 1666 he exhibited a model of the comprehensive rebuilding of the City, and was appointed a city surveyor to supervise much of Wren's work; he also designed a number of buildings, including the Royal College of Physicians and the Bedlam lunatic asylum. This post brought him several thousand pounds, which he kept untouched in an iron-bound box throughout his life, continuing to live in wretched conditions and dying meanly, without having tampered with microscopy again. Hooke is given undisputed credit for many innovations such as a hearing aid, an anemometer, a weather clock, a reflecting quadrant, a gear cutter for watch wheels, the wheel barometer, the construction of the first Gregorian telescope, the anchor escapement of clocks, and that small, "obvious", indispensable item of equipment, the universal joint.

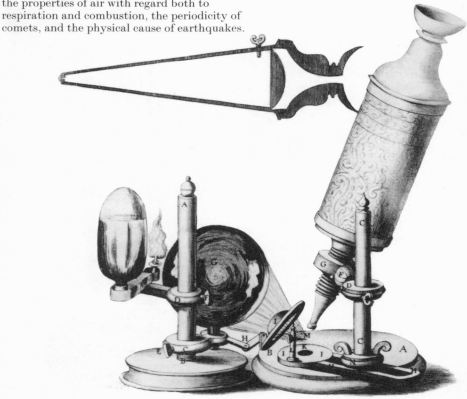

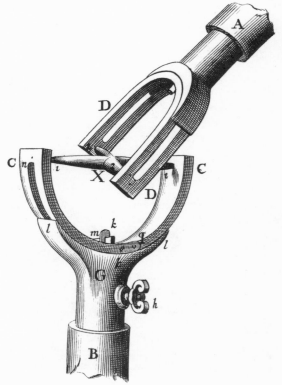

During his 10 years' attention to microscopy, Hooke had developed the compound microscope, which gave a magnification of up to 100 but, using two lenses, was imprecise in clarity. He had a near-contemporary in Holland – born three years earlier and dying, in his 91st year, 20 years later – who confined himself to a simple microscope utilizing one bi-convex lens through which the object was scrutinized after having been mounted on a vertical pin. This specialist was Anton van Leeuwenhoek of Delft (he is commonly identified with his home town, and seems never to have left it, existing on the profits of his drapery shop and a sinecure as janitor of Delft City Hall). Leeuwenhoek ground his own lenses, sometimes of minute dimension, with fine precision, obtaining magnifications of up to 150. He concentrated his observations on biological specimens, from fleas to spermatozoa. He explored blood cells and he discovered bacteria, though nothing was done about bacteria for a century because no one else had the ability to examine them in that interval. With Dutch as his only language, Leeuwenhoek could study the pictures in Hooke's *Micrographia* but not read the English words. Yet he corresponded at length with the Royal Society and was elected a fellow in 1680. With a skill as marked as any astronomer's he had seen farther *inwards* than anyone else of his age, and his accurate descriptions were the grist of invaluable medical and biological research.

Once the greater magnification of compound microscopes had been introduced, improvements were made in their precision. Until then, focusing had been done with sliding tubes but it was now given the greater control of a screw barrel by Giuseppi Campani, who had already perfected his lens-grinding lathe in Rome. In Campani's instrument, which was developed at about the time of the publication of Hooke's *Micrographia*, the tubes were made from hard wood on which coarse threads had been turned. Separated from the other scientist by 1,000 miles, Campani had to wait until he had read Hooke's work before learning that Hooke had modified the coarse adjustment of his sliding tube with a fine adjustment from a screw thread on the snout. This principle was extended in the series of telescopes manufactured by John Marshall from the end of the 17th century, in which the fine adjustment used a screw devised by the astronomer Helvelius of Danzig. Marshall made regular use of the ball-and-socket universal joint developed from Hooke's original design, so that the instrument could be turned to use available light to the best advantage.

Newton said, *If I have seen further than other men, it is because I stood on the shoulders of giants.* There were others in his time who, if not gigantic, were lofty enough to help each other over obstacles, though human enough to resent not getting their credits. Denis Papin graduated in medicine at Angers in 1669. But he devoted himself to physics and mechanics and soon went to the laboratory of the Académie des Sciences in Paris as assistant to his fellow Protestant Christiaan Huygens of Holland. Huygens was already experimenting with air pumps and, together with Papin, improved on Boyle's design. Leibniz introduced Papin to Boyle and by 1675, when he was 28, Papin was an aide to Boyle. But Papin individually developed his steam pressure cooker, introduced as *A new Digester or Engine for softening Bones.* He demonstrated it to the Royal Society in 1679, cooking in it a supper of fish, flesh, birds, and game, which gained him election as a fellow of the Society in the following year. He rejoined Huygens in Paris, later went to Venice, but came back to the Royal Society as paid curator exhibiting their experiments until he went on to professorships at Marburg and Cassel.

All through his peripatetic career Papin was concerned with the power of steam. The story of James Watt and the kettle should really be backdated by 90 years to embrace Papin observing the lid of his pressure cooker (which worked at three times the pressure of today's models) being forced upwards (he fashioned a special safety valve to contain it). Soon he was claiming that steam could be channeled to drive a paddle wheel in a ship, though he could not actually build a small boat working by a paddle until 1707. (It was destroyed by Luddite professional oarsmen.) In 1687 he designed a steam engine using a cylinder and piston, which was dependent on the creation of an efficient vacuum. But Thomas Savery built the first working steam engine 11 years later. Leibniz sent Papin a sketch of it.
Papin claimed to have improved on the design and pestered the Royal Society to conduct comparative tests. He died in poverty in London with his request unfulfilled. The year was 1712. Savery's engine had been further developed by Newcomen, and was being installed for practical industrial purposes.

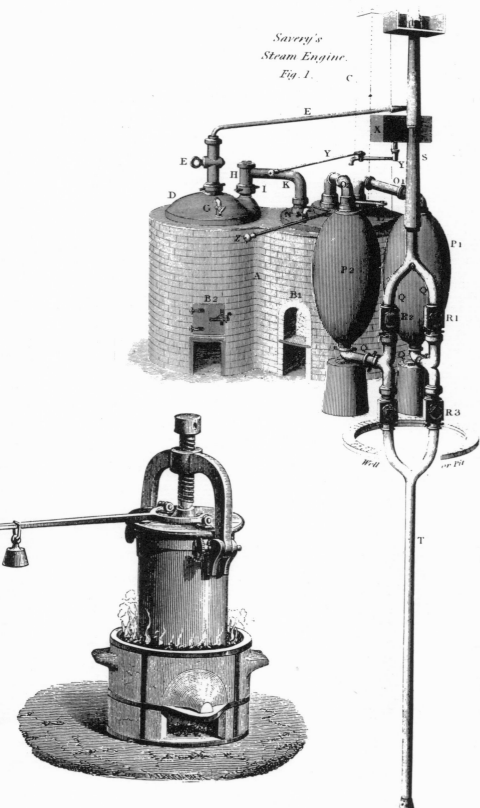

Savery's Steam Engine. Fig. 1.

The ENGINE Working in a MINE.

Thomas Savery, like Otto von Guericke, began professional life as a military engineer. He was a fair linguist and familiar with the published work of Guericke and Papin. Retiring to civilian life, where he always called himself Captain, he set himself to tackle one of the most pressing industrial problems of the times: how to pump out the water which continually seeped into mines. Mines were getting deeper and horse-driven pumps were becoming uneconomical even when space could be found for them. In 1698 Savery devised what he called *The Miner's Friend*: a pump actuated by the alternation of an induced vacuum and high-pressure steam. A boiler produced steam which was introduced into a 50-gallon vessel and then cooled by dousing the outside of the vessel with cold water. The resultant vacuum drew water into the vessel from the lower level which was to be emptied. A further injection of steam under pressure forced the water further up the pipe. A twin vessel duplicated the action in series, one being filled while the other was emptied. In practice, the vacuum lifted the water only 20 feet (because of the weight of the atmosphere, recently established by Torricelli), and the existing apparatus accommodated high-pressure steam only with difficulty. Savery's engine befriended very few miners, but was used for some decades to pump water to the upper stories of country houses.

Savery came from the hill country in north Devon. Farther south in Devon a Dartmouth blacksmith, Thomas Newcomen, was also familiar with the need for efficient pumping, particularly in the Cornish tin mines. He devised a pump that worked by the introduction of steam into a cylinder. After the steam had been cooled by an *internal* spray of water, a piston was driven down by atmospheric weight on the partial vacuum, rather than a draft of high-pressure steam. It was a true reciprocating engine. Newcomen found that the patent taken by Savery covered his conception, and he therefore went into partnership with Savery. The Newcomen engine was low-powered, developing some five horsepower; it was slow, with 14 strokes to the minute; but though it was not of ideal mechanical efficiency, it was effective, for it was constructed with the highest precision then available and with the solidity of the newly conquered Rock of Gibraltar. It was first installed in a mine near Dudley in Staffordshire as Papin was dying obscurely in London. The Newcomen engine was widely used in Europe and America throughout the 18th century.

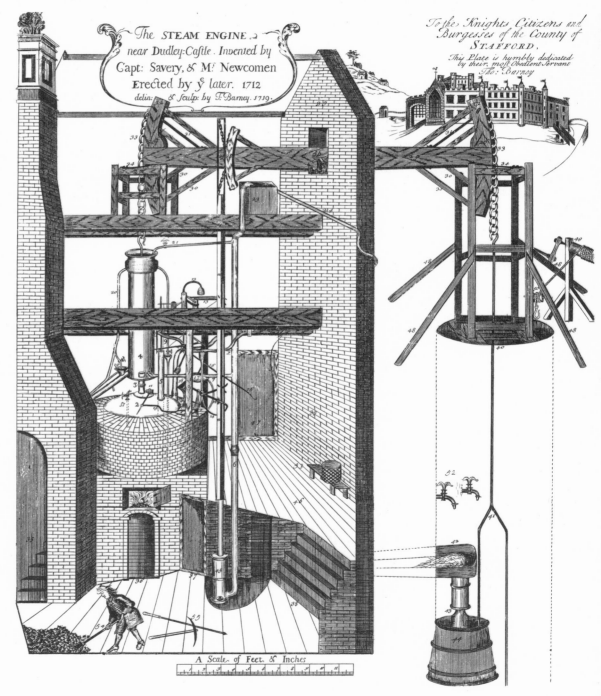

The STEAM ENGINE near Dudley-Castle. Invented by Capt: Savery, & M.ʳ Newcomen Erected by ẙ later. 1712 delin: & sculp: by T. Barney. 1719.

To the Knights Citizens and Burgesses of the County of STAFFORD. This Plate is humbly dedicated by their most Obedient Servant Tho: Barney

A Scale of Feet & Inches

With the enormous surge in overseas trade in the 17th century, particularly between Britain and America and often under conditions of war, accurate navigation became a matter of supreme commercial importance.

This development had particular significance for the British who, in the words of Adam Smith, were intent on founding *a great empire for the sole purpose of raising up a people of customers (which) may at first sight appear a project fit only for a nation of shopkeepers*. With the expansion of the American trade and the establishment of a naval dockyard at Plymouth, the gale-swept location of the Eddystone rocks became one of the most menacing areas in British waters. Located 14 miles south-south-west of Plymouth, Eddystone had a triple underwater reef, of which only the center ridge was permanently exposed, and increasingly the cause of shipwrecks. No one knew this better than the entrepreneur and engineer, Henry Winstanley, who had lost one of his ships there. In 1696, at age 44, Winstanley took up a contract to build a much-needed lighthouse on the Eddystone. On a solid drum of masonry, finally 20 feet in diameter and 18 feet high, he erected a polygonal tower 56 feet high topped by an octagonal lantern containing a candelabrum. While he was supervising the building, Winstanley was hijacked by a French privateer, but was graciously returned by Louis XIV, who declared that *he was at war with England, not with humanity*.

The Eddystone light first shone on November 14, 1698. Winstanley used to visit it regularly. He was in the lighthouse on the night of November 26, 1703, when the entire tower was washed away in a historic storm, from which no bodies were recovered. John Rudyerd built a second Eddystone lighthouse of laminated timber on masonry, which was in operation from 1708 until 1755, when it was destroyed by fire. Robert Smeaton's tapering tower entirely in masonry lasted from 1759 until its replacement in 1882 by the graceful tower of James Douglass.

Since there are no landmarks in mid-ocean the navigator must have an appreciation of latitude – the angle between the plumb line of a particular location and the plane of the equator. He must also know longitude – the angle which the terrestrial meridian connecting the location with the poles makes with the standard meridian, now universally accepted as at Greenwich. Determination of both latitude and longitude demands celestial observation, either of the stars or of the sun. Angles must be taken in the same manner that astronomers take angles. The first reliable instrument to measure angles in the sky was Hooke's octant. The one application for which it had limited use was on the unstable levels of a ship at sea – as Isaac Newton pointed out to Halley. Newton delineated the requirements of a nautical instrument to determine altitudes. (It is called *octant*, *quadrant*, or *sextant* according to whether it is based on an eighth, a quarter, or a sixth of the circumference of a circle.) These requirements were fulfilled in 1730 by the reflecting quadrant of James Hadley, a Hertfordshire mathematician and instrument maker. Hadley's quadrant was designed for use at sea, to measure the angle between two observed objects. Manipulation of the quadrant brought images of the objects, held in two small mirrors, to the point where they coincided. (Hadley's quadrant was in fact an octant, but the double reflection qualified it as measuring up to 90 degrees. When it was extended in 1757 to take in the arc of one-sixth of a circle and an angle of 120 degrees, it was renamed the sextant.) In 1734 Hadley had fitted a spirit level to his quadrant so that a meridian altitude could be taken at sea in conditions when the horizon was not visible.

The problem of longitude had for years concerned the maritime powers. The Dutch and Spanish governments had offered prizes for its solution, which Galileo vainly claimed in 1616 by tabulating the predictable eclipses of satellites of Jupiter, which he had discovered in 1610.

The Royal Observatory was set up by King Charles II at Greenwich in 1675 precisely *for rectifying the tables of the motions of the heavens and the places of the fixed stars so as to find out the so-much desired longitude of places for perfecting the art of navigation.*

It was because of this early foundation that the meridian crossing Greenwich was slowly adopted as zero for the world. However, celestial observations, whether of Jupiter or of the Moon (the latter being strongly backed by British astronomers royal), were unreliable for a century because of the inaccuracy of the nautical tables. The alternative lay with the recognition of longitude as a version of the measurement of time, and the construction of accurate chronometers which could be set up at Greenwich and would work reliably afterwards. In 1714 the Board of Longitude was established as a department of the British Admiralty, jealously supervised by the Royal Observatory, with immediate power to offer prizes of up to £20,000 for the construction of a timekeeper which could determine longitude with an accuracy of up to 30 miles. John Harrison, a Yorkshire carpenter and joiner who in 1715, at the age of 22, had built a practical eight-day clock entirely of wood, was determined to win the prize. In 1736 he built the first of five successive timekeepers. All incorporated Harrison's use of steel bound with brass in the pendulum or the balance (the wheel which spins back and forth), so that their differing expansion rates maintained regularity during changes of temperature. The first fitted into a two-foot cube and weighed 72 pounds.

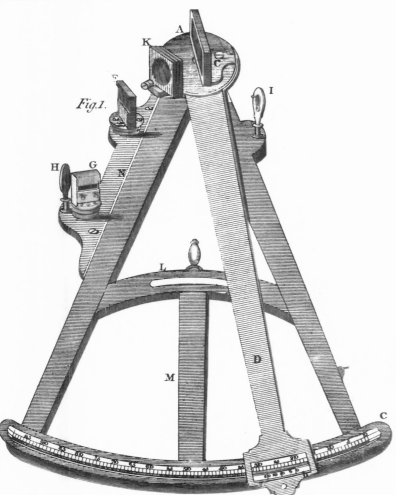

Fig.1.

Textile working by man stood remarkably still for 8,000 years after its inception in about 6500 BC, when it began with the processing of wool, moving on to work flax fibers into linen, and soon to the efficient manufacture of cotton and silk. Spinning, the making of a long continuous thread by twisting short natural fibers under tension from the human hand, hardly changed through the millennia until the spinning wheel was introduced in the 14th century AD. Weaving, using a loom as a frame to separate alternate warp threads so that the weft thread could be passed through in a shuttle to interlace them, did not change in principle for an even longer period. In England in the 18th century there was a succession of fundamental changes in textile-making methods, facilitated by the fact that, as a former cottage industry, textile manufacture was not fossilized by restrictions operating in the guilds controlling more organized crafts, but could respond more fluidly to the incentive for cheaper, faster production.

The first changes in English textile machinery in the 18th century were not concerned with the adaptation of power but with the application of ingenuity. In 1730, when the 26-year-old John Kay of Bury was dickering with improved loom-reeds and carding appliances, the only available *power* – apart from the traditional trio of muscle, wind, and water – was Newcomen's steam engine, and at that time its only practical use was to pump. Conversion to rotary movement, which would make the steam engine the first versatile prime mover, was half a century away.

Kay's important contribution was little more than the exploitation of a dexterous twist from a muscular wrist.

In 1733 Kay evolved the flying shuttle. The shuttle holding the weft thread in the weaving loom was shot through the warp threads by the flick of a cord controlling a shuttle box at either side. Adam Smith said that the previous invention of the spinning wheel and flyer doubled the speed of hand-spinning. Kay doubled the speed of hand-weaving and halved the manpower necessary for weaving broad pieces. Until then, the shuttle was manually thrown and a single weaver could not handle a piece wider than the reach of both hands – in practice, 32 inches. With the flying shuttle the work was not only faster and cheaper, but the product was improved, since the weaver had a free hand to beat the weft into the cloth more skilfully.

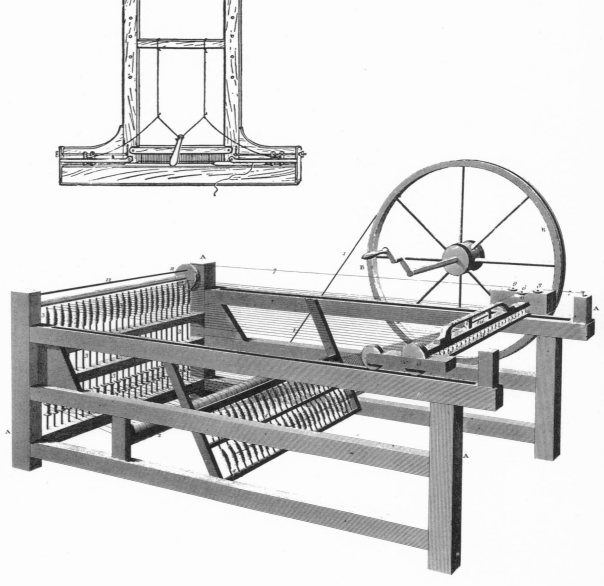

Kay's fly shuttle was not speedily adopted, though it represented such a threat to weavers that a mob in Bury burned his house and forced him abroad. Even where the system was adopted, the manufacturers refused to pay royalties and Kay could not afford to sue. Meanwhile, John Wyatt of Lichfield, in partnership with Louis Paul, developed a machine capable of spinning more than one thread at a time, with rollers revolving at progressively faster speed to control tension. Their machines were installed in workshops between 1740 and 1744. But their lack of impact was signified by the prizes awarded by the Royal Society of Arts in 1761 for the invention of a machine that would spin six threads at once when worked by only one operator. In 1764, Kay of Bury died poverty-stricken in France while his fly shuttle was coming into general adoption in England. The same year, James Hargreaves developed the spinning jenny, which could spin eight threads at one time and – ominously – could be more easily worked by children than by adults. Hargreaves was a handloom weaver, and at first kept the jenny in his home, where his children produced weft for his own weaving. Yet within four years, as he began to sell his machine to manufacturers – who increased the number of spindles from eight to 120 and again dodged royalties – the home spinners of his native Blackburn gutted his house and destroyed both his jenny and his loom.

In 1767 Richard Arkwright, a 31-year-old Lancashire entrepreneur, engaged John Kay of Warrington, a clockmaker, to lend his skill in making a new spinning machine. Their spinning frame, patented in 1769, was a most original arrangement of rollers and spindles, spinning yarn of a much higher quality – stronger, smoother, more regular – than the spinning jenny. Significantly, however, whereas the jenny was small enough to be used domestically, the spinning frame demanded power. Arkwright promptly equipped a factory in Nottingham to spin the new yarn.

For power he used a water-wheel driven by the river Trent, copying for this first spinning mill the only practical model in England, a silk mill in Derby erected after the secret technology of the 17th Century silk mills in Italy had been expounded by Vittorio Zonca in his *Novo Teatro di Macchine*. Arkwright later perfected a carding engine and other refinements which combined every process of yarn manufacture when they were incorporated into what continued to be called the *water frame* even after it was converted to steam power.

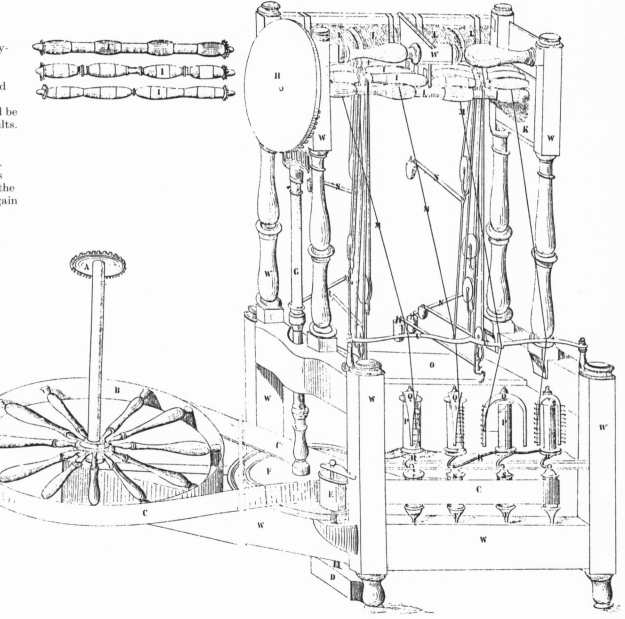

153

Between 1775 and 1779 Samuel Crompton, a fiddler in a Bolton theater, spent his free time trying to improve the spinning jenny, at which he had worked during his orphaned childhood. He produced a "mule," a hybrid between the jenny and the water-frame principle, capable of spinning yarn fine enough for the production of delicate muslins which previously had been imported from India. His particular innovation was the spindle carriage. Moved by the operator's hand and knee, it controlled the tension as the roving sliver was elongated and twisted. This action faithfully duplicated that of the human hand in spinning fine yarn, but on a multiple scale. With Crompton's mule the mechanization of spinning was completed. By 1811 there were four-and-a-half million spindles being operated on mules, 10 times as many as on jennies and water frames. Power was introduced to weaving in 1785 by Edmund Cartwright's power loom, but progress in this area of weaving was comparatively slow until Richard Roberts created precision-made power looms between 1822 and 1830: handloom weaving was common into the second half of the 19th century. It was in the *dark Satanic mills* of the spinning masters that the factory system was fully developed, and Arkwright, with his flair for organizing capital, enthusiastically took on the role of Lucifer, in spite of the Luddite fury which hit him and Crompton.

Rigorous discipline, 12-hour labor shifts night and day, a ruthless dismissal policy, and the full exploitation of child labor enabled him to achieve success in what his kindly Victorian biographer described as *overcoming the prejudices of the workers, accustoming them to unremitting diligence during the stated hours of labour, training them for their particular tasks and inducing them to conform to the regular celerity of the machinery.* Although he lost many legal suits claiming unpaid patent rights, Arkwright was alone among the textile innovators in making any substantial money. When he died in 1792 as Sir Richard Arkwright he left half a million pounds, and his son began the pleasant habit of presenting his family with £10,000 notes wrapped in table napkins on Christmas morning.

Early cotton machinery, Sir John Hicks has said, *fits better as an appendage to the evolution of the old industry than in the way it is usually presented as the beginning of the new . . .* There is continuity between the eighteenth century development of Lancashire and the West Riding and the pre-Industrial Revolution world. There might have been no Crompton and Arkwright and still there could have been an Industrial Revolution.

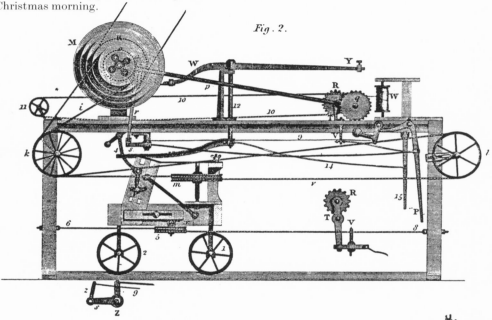

Fig. 2.

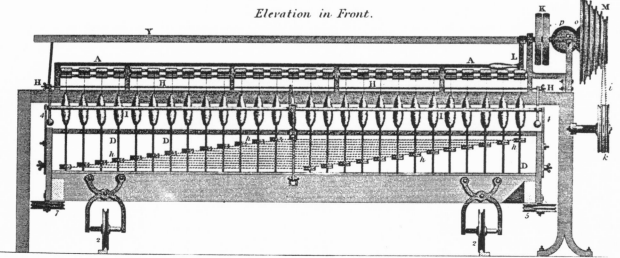

Elevation in Front.

THE INDUSTRIAL REVOLUTION

Between 1780 and 1850, in the space of less than three generations, the face of England was changed by a far-reaching revolution, without precedent in the history of mankind. From then on, the world was no longer the same.

The animal and vegetable kingdoms had traditionally provided most of the energy man needed to keep himself alive, propagate the species and make a living. The Industrial Revolution opened up a completely different world of new and untapped sources of energy such as coal, oil, electricity and the atom, exploited by means of various mechanisms – a world in which man found himself able to handle great masses of energy to an extent inconceivable in the preceding bucolic age. From a narrow technological and economic point of view, the Industrial Revolution can be defined as the process by which a society gained control of vast sources of inanimate energy; but such a definition does not do justice to this phenomenon, neither in terms of its distant origins nor in its economic, cultural, social and political implications.

Despite great upheavals such as the rise and fall of the Roman Empire, Islam and the Chinese dynasties, a basic continuity characterized the pre-industrial world. Agronomists in the 15th and 16th centuries could still usefully refer to treatises written by the Romans. The ideas of Hippocrates and Galen continued to represent the basis of official medicine well into the 18th century. It did not seem absurd to Machiavelli to refer to the Roman constitution when he planned an army for his times. Palladio and his successors could still draw inspiration and instruction from the buildings of classical antiquity. If a Roman of antiquity could have been transported some

Overleaf: Edmund Lee's fantail windmill, patented 1745

18 centuries forward in time, he would have found himself in a society which he could have learned to comprehend.

This continuity was broken between 1750 and 1850. In the middle of the 19th century, if a general studied the organization of the Roman army, if a doctor paid attention to Hippocrates or Galen, or if an agronomist read Columella, he did it purely for historical interest and as an academic exercise. Even in faraway, unchanging China the most enlightened of the scholar-administrators of the Celestial Empire began to realize that the ancient classical values which had given continuity to Chinese history were no longer valid for survival in the contemporary world. In 1850 the past was not merely past – it was dead.

On the other hand, while the Industrial Revolution created an irrevocable discontinuity in the course of history, its roots nevertheless reached deep into the preceding centuries. Its origins can be traced to the profound change in ideas and social structures that accompanied the rise of the urban communes in Northern Italy, in Northern France and in the Southern Low Countries between the 11th and the 13th centuries. Then the warlord and the monk were replaced by the merchant and the professional. Within a few centuries, the new civilization had conquered Western Europe, changing both institutional and human structures. In the 16th and 17th centuries severe crises hit Italy and the Southern Low Countries; but the movement along commercial capitalistic lines reached its height in the Northern Low Countries and in England. At the end of the 17th century the outstanding features in these two areas were an extraordinary expansion in commerce and manufacturing: the emergence of a large merchant class endowed with remarkable managerial ability, economic power and social and political influence; an impressive stock of manpower, both artisans and professionals, and a relative abundance of capital. These were the material facts. In the realm of ideas, the main characteristic was a relatively high degree of literacy, an aptitude for things mechanical, and a strong and growing inclination towards quantitative measurement and experiment.

By the end of the 17th century, Holland, not England, would have seemed to be the one country best suited to an explosive revolution in the field of production. But Holland was imperceptibly ossifying into conservatism, and she was losing leadership in a number of fields. Moreover, England possessed coal and Holland did not.

The use of coal for heating and iron ore smelting in England went back to the 16th century, and abundant deposits of coal almost at surface level facilitated progress in the technique of its use. Toward the end of the 18th century, Watt's steam engine made possible the transformation of the chemical energy of coal into mechanical energy. After 1820, when the steam engine was used for rail transport, coal became available for widely varied processes of production. The world production of coal and equivalent energy developed in a spectacular way.

Watt's discovery was no accident. Nor was it an accident that the discovery had an overwhelming application in the field of production, and was followed by a whole series of similar inventions. Man *had invented the method of invention*, and new discoveries would allow the more efficient use of the already known forms of energy and the exploitation of new forms. Between 1860 and 1890 the industry of oil extraction was begun, and the internal combustion engine was brought to perfection. The end of the century saw the introduction of electricity. In the middle of the 20th century man had begun to exploit atomic energy. World production of coal continued to increase but its share of the total energy derived from inanimate sources was continually decreasing.

The availability of coal was a strategically important element in the process of development until the middle of the 19th century. However, if coal was a necessary condition for industrial growth it was never a sufficient condition. An industrial revolution is above all a social and cultural development. Coal by itself does not create and does not move machines. There must be men capable of mining the coal; of devising, building and operating the steam engines; of organizing the factors of production and assuming the risks and responsibilities of enterprise. From the middle of the 19th century onward the reduction in the cost of transportation of coal and the economic exploitation of alternative types of energy made industrialization possible in those areas which were not endowed with coal fields. This greatly helped the spread of the Industrial Revolution. But the sociocultural roots of the Industrial Revolution become evident when one notices that the first countries to industrialize were those with the highest literacy rates and the greatest cultural similarities with England.

The spectacular growth in industrial output was accompanied by a spectacular growth in population in such countries as England, France, Germany, Switzerland and the United States. The population of England and Wales rose from about six million in 1750 to about nine million in 1800 and about 18 million in 1850. Between 1750 and 1850 the population of Europe (Russia excluded) rose from about 120 million to 210 million. In 1950 it reached 393 million. Amazing as it was, the growth in population was still lower than the growth in national product. In the 10 years before 1780 income per head in England increased by about 0.3 percent per annum. Between 1850 and 1900 it increased by more than two percent per annum.

The meaning and consequences of this growth in income per head are made obvious by other statistics. In a pre-industrialized country, life expectancy is less than 30 years, more than half the average personal income is absorbed by the cost of living, and the same income cannot assure subsistence during a time of famine. In an industrialized country, food expenditure absorbs no more than a quarter of the average personal expenditure, and life expectancy is well over 60 years. No matter what index one selects, a dramatic change in the figures takes place corresponding to the Industrial Revolution. It was a revolution without precedent in history: a leap forward into a completely new world.

All the 17th-century pioneers up to Savery used steam, or the vacuum created by condensing steam, to work on the medium they were actually manipulating – generally water that had to be shifted in a shaft. Newcomen made a real steam engine, with moving parts worked by steam, which transferred their motion to work other machinery, principally pumps. James Watt made Newcomen's engine more economical and more powerful, with automatic control of its speed, and he made it turn a wheel. Since Watt, rotation has been the first property of power.

At the age of 21 Watt was a fledgling mathematical instrument maker at the University of Glasgow. Seven years later, in 1764, he was concerned by the poor performance of a Newcomen engine which had been overhauled by the most skilful instrument maker in London. He analyzed its basic inefficiency. After each stroke the steam within the cylinder had been condensed by cold water. Steam had to be reintroduced for the next stroke, and had to re-heat the whole of the chamber: cylinder walls, piston and its packing. Watt calculated that an efficient engine would require only one third of the steam (and therefore fuel) of the Newcomen type.

Watt decided to keep a cylinder at constant high temperature and to produce a separate condenser, connected to the cylinder and constantly cooled, in which condensation was to take place. After experimenting with a condenser cooled only by the cold water in which it was immersed, he reintroduced the principle of condensation by a jet of cold water and added an air pump which ejected that water, with the condensed steam and much of the air itself, so that the chamber was in a state of partial vacuum.

This was the first Watt steam engine, patented in 1769 but not in production for sale until 1775. It was much faster than the Newcomen engine it replaced, since there was no pause for reheating the cylinder, and it was two to three times cheaper in fuel costs.

Mʀ WATT'S ENGINE.

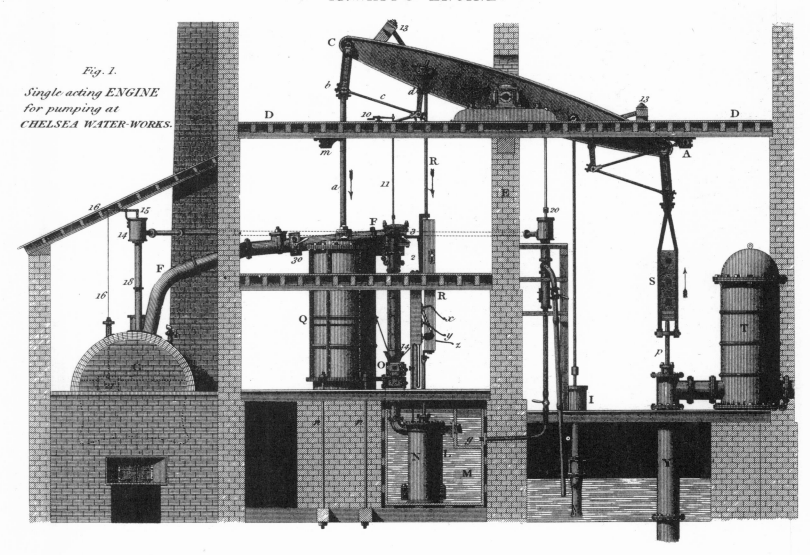

Fig. 1.

Single acting ENGINE
for pumping at
CHELSEA WATER-WORKS.

In the patent of 1769 Watt had included specifications for making the engine drive a wheel, though in fact his first design was impractical. Jonathan Hull's undeveloped patent of 1736 envisaged a tugboat powered by a Newcomen engine operating paddle wheels. As early as 1770, some 30 years before the first use of the screw propeller and 70 years before its successful application, Watt proposed a boat that could be driven by a steam engine operating a *spiral oar*, and sketched a screw propeller. But the principle of rotation, first used in water transport to drive paddles, was really dependent on Watt's later introduction of the double-acting engine.

The first single-acting rotary engine in a self-propelled vehicle was created in 1770 by Nicolas Joseph Cugnot, an engineer in the French army, when he devised a steam carriage to haul artillery into action. The Mark II model of 1771, which still exists, is a three-wheeler with a heavy copper boiler positioned ahead of the single front wheel. Over the wheel are slung two 50-liter, single-acting, high-pressure steam cylinders driving ratchets on either side of the axle, and developing a speed of two to six miles an hour with four riders up.

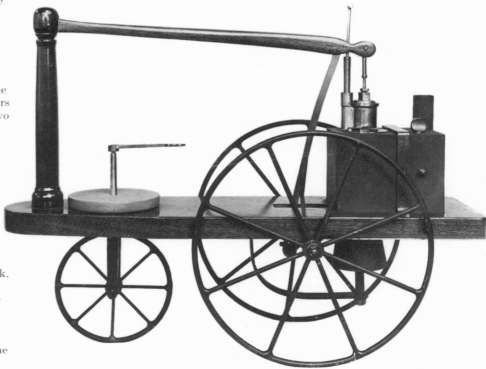

Not until 1784 did Watt patent a steam carriage for public highways, using a differential gear for hills, and Watt's assistant William Murdock, working with Cornish mine owners who were installing B & W engines, built a model which he ran on a Cornish road the same year. The three-wheeled steam car ran away at such speed that it convinced the local parson it was a manifestation of the Devil. Murdock improved Watt's design but could not exploit it, for Watt – cautiously, or possibly jealously – assigned to his assistant a suffocating increase in work on the improvement of the stationary engine.

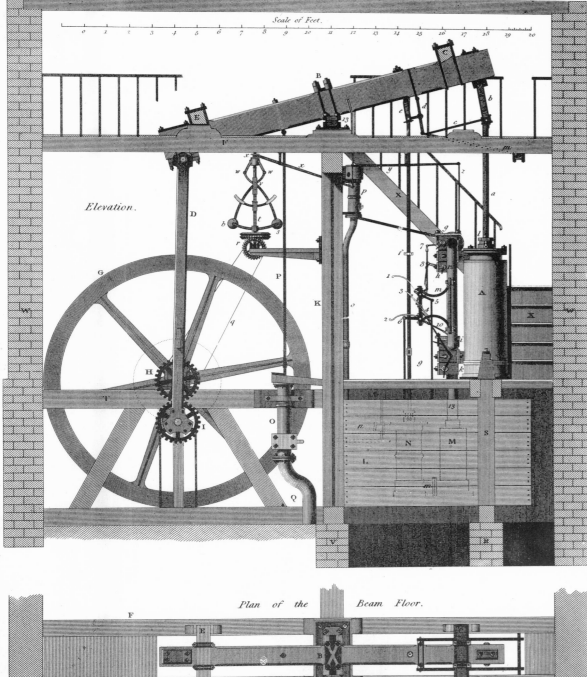

Scale of Feet.
0 1 2 3 4 5 6 7 8 9 10 11 12 13 14 15 16 17 18 19 20

Elevation.

Plan of the Beam Floor.

Watt had developed the double-acting engine by 1780, introducing steam and a subsequent vacuum on *both* sides of the piston alternately. But the timid inventor had been forced out of utilizing a crank to obtain rotary action because one of his workmen maliciously patented the system on which he knew Watt was working. Watt therefore produced the "sun-and-planet" system (the name is a telling reminder of how much astronomy influenced practical scientific thought). The engine drove the rod connecting to the "planet" cogwheel. which gave the "sun" wheel shaft (also the shaft of the flywheel) two revolutions for each of the engine's revolutions, doubling its effective speed. Finally Watt adopted a self-regulating speed governor, which flung centrifugally rotating balls higher as the speed of an engine increased; its scissor-like move-ment pulled down a balanced bar controlling a throttle valve, which decreased the supply of steam into the engine; and its *vacillation* produced a constant speed. This was not a new principle in engineering but it had not previously been used to control speed. From 1785, two years before he officially introduced the governor, Watt had been quietly testing his device in congenial surroundings – on Boulton & Watt rotative engines installed in Whitbread's brewery in Chiswell Street, London. The shy and unbusinesslike Watt might never have produced his engine without Matthew Boulton, a Birmingham metal manufacturer who gave him constant moral support and paid his debts for patenting and development in return for two-thirds of the profits. Boulton had the patent extended until 1800, by which time the firm of Boulton & Watt sold some 500 engines to enterprising men such as Richard Arkwright.

PLAN
of the Great Wheel.

Fine precision in metalworking had depended on the lathe since its practical development in the 16th century. Until the Industrial Revolution such work was mainly required for scientific instruments, particularly with screws which served as holding devices and as the most reliable way to make delicate adjustments on clocks and on instruments used for measurement, microscopy, and astronomy. Jesse Ramsden, the son of a Halifax innkeeper, entered the field late, when he began his apprenticeship at age 23 as a mathematical instrument maker in London in 1758. Shortly after setting up on his own in 1762, he married the daughter of John Dollond, FRS, and gained as his dowry a share in Dollond's patent for making achromatic lenses. At his workshops in the Haymarket and Piccadilly Ramsden was concerned not only with telescopes and mathematical instruments: he had acquired such skill as an engraver that many artists from the newly established Royal Academy commissioned him to make plates from their work. In 1763 Ramsden made a "dividing engine" for the accurate subdivision and graduation of circles which he used to make more accurate mathematical instruments. Eventually he employed 60 craftsmen to produce instruments still prized for their workmanship. One of his specialties was making the clockwork movements for observatory telescopes. He also made more than 1,000 sextants, incorporating improvements on Hadley's design. A by-product of his urge for accuracy was his development of the precision screw-cutting lathe, a fine machine only 12 inches long. This was perfected in the 1770s.

In the same period a heavier industrial lathe was being developed in France by the veteran Jacques de Vaucanson, who built into it a sliding tool carriage advanced by a long leading screw. Vaucanson became a specialist in lathes and drills because of the precision needed for his early interest in the construction of automatic toys: a characteristic automaton was his model duck which quacked and flapped its wings, swam, and even swallowed food and "digested" it in a sort of mincer. In 1745, at the age of 36, when he was a government inspector of silk factories, he had made a novel self-acting loom which did not, however, include a fly shuttle; and he made the first self-acting loom capable of being driven by power. Ramsden and Vaucanson were pioneers in applying to the craft of metalworking the cutting and shaping of a revolving workpiece, which had previously been restricted to the woodworking lathe.

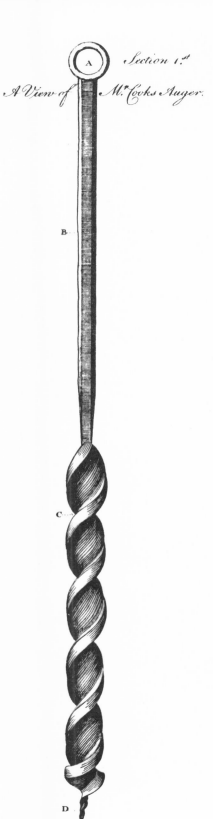

The spiral auger, a boring tool with an Archimedes screw to lift the debris from the hole being bored, made an astonishingly late appearance in 1770. Phineas Cooke demonstrated it during that year to the learned Society for the Encouragement of Arts, Manufactures and Commerce (now the Royal Society of Arts). The society's registrar, William Bailey, described in somewhat naive wonder its six-inch working part below the six-and-a-half-inch shank: *An endless screw, with a double worm or thread a quarter of an inch thick, and flat on their edges; these worms cut two spiral chips which pass through the two concave spiral channels of the auger, and are gradually discharged therefrom without drawing out the auger till it has bored a hole full three feet deep, or any other depth required. The point of the auger is a taper screw with a double worm, like a gimblet, which pierces the wood much easier and truer than common augers, and requires no picking with a gouge, which in the usual method is an unavoidable operation attended with a great deal of trouble and loss of time.* The Society's Committee of Mechanics saw Cooke's auger as an instrument promising to be *a tool of great use in shipbuilding, &c.*, and recommended a bounty of 30 guineas payable on *his leaving the instrument with the Society for the use of the public.*

Accurate boring of material far more intransigent than ships' bottoms was necessary when James Watt faced the problem of constructing a cylinder for his steam engine that would not leak steam as the piston ran in it. Matthew Boulton had told him in 1769 that he had need of *as great a difference of accuracy as there is between the blacksmith and the mathematical instrument maker.* When Boulton became the partner of Watt, they took the problem to John Wilkinson, the English Midlands ironmaster. Wilkinson had recently, in January 1774, developed a cannon-boring mill in which the solid casting of the barrel was rotated horizontally while the stationary boring head was advanced by a toothed rack on the boring bar, which was well supported. In 1775 Wilkinson adapted this cannon borer to produce for Boulton & Watt a cast cylinder 50 inches long and 18 inches in diameter, hollow, with the boring bar running through the cylinder and supported at both ends. The accuracy of this essential job, Boulton ecstatically declared, *doth not err the thickness of an old shilling* – about 1/1,000 of an inch per inch of diameter. This feat enabled Watt to produce his steam engine in 1775. The first boring was done by the power of a water wheel. Wilkinson became a priority customer for steam engines to power his machines, a blast furnace, a primitive power hammer, and a rolling mill. Wilkinson went on to bore cannon for Wellington in the Peninsular War, though he also smuggled matching artillery to the French who were fighting Wellington.

John Harrison had continued to make a succession of chronometers after the refusal of the Board of Longitude to reward him for his 1736 timekeeper. His fourth model, reduced to the size of a pair-case watch five inches in diameter, has been nominated by Lieutenant Commander Rupert Gould, navigator and horologist extraordinary, as *by reason alike of its beauty, its accuracy, and its historical interest . . . the most famous chronometer that ever has been or ever will be made*. The silver-cased watch has a white enamel dial ornamented in black, with hour and minute hands of blued steel and a very early example of a center second hand. Although it was designed as a marine chronometer, Harrison made no provision for suspending it steady in gimbals, which in his experience caused more errors than they eliminated. Instead, he laid it in a box on a soft cushion, its level slightly tilted from the horizontal. Harrison's "No. 4", made in 1759, was tested two years later on a voyage to the West Indies and was found to be only five seconds slow, an error of about one mile in establishing a landfall. The Board of Longitude refused to pay the prize and another test was arranged. Matters dragged on for 11 years until King George III assured the 79-year-old craftsman *By God, Harrison, I'll see you righted*, and payment of the reward was completed in 1772. This was not the only injustice exercised by the quirky Board of Longitude. At the time of the King's intervention it had appointed Joseph Priestley, a Unitarian minister, as scientific adviser to Captain James Cook, who was to sail to the South Seas to undertake marine surveying, astronomical observation, and the search for Australia. In the words of Samuel Smiles, *Dr Priestley's appointment was cancelled since the Board of Longitude objected to his theology*.

Part of the Board of Longitude's ploy for delaying payment was to order Harrison to take his "H 4" timekeeper apart in front of six men, who included the watchmakers Larcum Kendall and Thomas Mudge. Kendall was then instructed to make an accurate copy of "H 4," which he finished in 1769. Captain Cook took the copy on his first three-year exploration and praised it highly. Kendall had also simplified "H 4" into his second watch, made for the Board of Longitude; a graceful timekeeper finished in 1772, it was known as "K 2". This, though never approaching the accuracy of its two predecessors, was severely tested under Polar conditions and was then allocated to Captain William Bligh for his voyage aboard the *Bounty* in 1787. The mutineers kept the chronometer when they set Bligh adrift, and took it to Pitcairn Island when they occupied it in 1790. In 1808 "K 2" was sold to the captain of the American whaler *Topaz*, who had called at Pitcairn looking for seals and fresh water, and thus discovered the hitherto unknown haven of the mutineers. Shortly afterwards, the watch was stolen. It was later resold in two ports in Chile and returned to England when it was 60 years old. Thomas Mudge, after inspecting Harrison's fragmentation of "H 4", was stimulated to spend the rest of his life trying to improve on the timekeeper. His only lasting achievement was to invent the lever escapement. Unrecognized at the time, it was almost universally adopted much later. Further development of the chronometer was achieved by two bitter rivals, John Arnold and Thomas Earnshaw, who produced between them some thousand marine timekeepers at comparatively low prices, using a rationalized allocation of labor among craftsmen-employees.

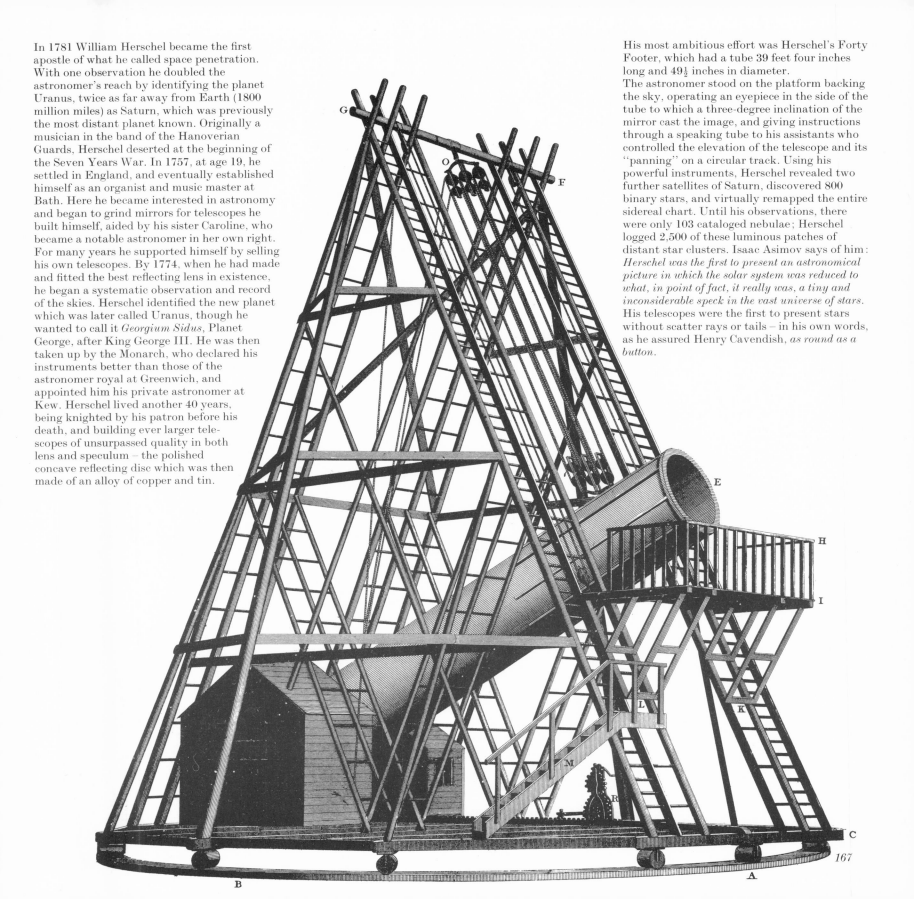

In 1781 William Herschel became the first apostle of what he called space penetration. With one observation he doubled the astronomer's reach by identifying the planet Uranus, twice as far away from Earth (1800 million miles) as Saturn, which was previously the most distant planet known. Originally a musician in the band of the Hanoverian Guards, Herschel deserted at the beginning of the Seven Years War. In 1757, at age 19, he settled in England, and eventually established himself as an organist and music master at Bath. Here he became interested in astronomy and began to grind mirrors for telescopes he built himself, aided by his sister Caroline, who became a notable astronomer in her own right. For many years he supported himself by selling his own telescopes. By 1774, when he had made and fitted the best reflecting lens in existence, he began a systematic observation and record of the skies. Herschel identified the new planet which was later called Uranus, though he wanted to call it *Georgium Sidus*, Planet George, after King George III. He was then taken up by the Monarch, who declared his instruments better than those of the astronomer royal at Greenwich, and appointed him his private astronomer at Kew. Herschel lived another 40 years, being knighted by his patron before his death, and building ever larger telescopes of unsurpassed quality in both lens and speculum – the polished concave reflecting disc which was then made of an alloy of copper and tin.

His most ambitious effort was Herschel's Forty Footer, which had a tube 39 feet four inches long and 49½ inches in diameter.

The astronomer stood on the platform backing the sky, operating an eyepiece in the side of the tube to which a three-degree inclination of the mirror cast the image, and giving instructions through a speaking tube to his assistants who controlled the elevation of the telescope and its "panning" on a circular track. Using his powerful instruments, Herschel revealed two further satellites of Saturn, discovered 800 binary stars, and virtually remapped the entire sidereal chart. Until his observations, there were only 103 cataloged nebulae; Herschel logged 2,500 of these luminous patches of distant star clusters. Isaac Asimov says of him: *Herschel was the first to present an astronomical picture in which the solar system was reduced to what, in point of fact, it really was, a tiny and inconsiderable speck in the vast universe of stars.* His telescopes were the first to present stars without scatter rays or tails – in his own words, as he assured Henry Cavendish, *as round as a button.*

On June 4, 1783, in the French market town of Annonay, near Lyon, two brothers, Joseph and Jacques Montgolfier, master papermakers aged 43 and 38, lit a pyre of wool and straw. They led the smoke into a linen container made airtight with a gummed paper lining, which filled to form a globe 11½ meters high. Tethering ropes were released and the contraption, which the Montgolfiers called a *ballon*, rose some 2,000 meters and drifted some two kilometers before landing. It was the first calculated balloon ascent. The Montgolfier brothers had long determined to *enclose a cloud in a bag*. They had failed with steam (which condensed too soon) and with hydrogen (impurely made, so that free sulphuric acid rotted the bag). The smoke which they finally used was not recognized as hot air, but for some time was termed *Montgolfier's gaz*.

The Montgolfiers were invited to Paris by the Académie des Sciences. They put animals into balloon flight and King Louis XVI, after striking a medal in their honor *pour avoir rendu l'air navigable*, suggested that two condemned criminals should go next. However, the privilege of initiating human flight was claimed by a 29-year-old chemist, Pilâtre de Rozier. With an infantry major, the Marquis d'Arlandes, he ascended from the royal Château de la Muette in the Bois de Boulogne on November 21, 1783, and travelled for 12 kilometers in a hot-air balloon of 2,200 cubic meters capacity. D'Arlandes stoked the brazier with straw, and swabbed out occasional fires in the envelope with a wet sponge. On June 15, 1785, de Rozier crashed to his death with P.A. de Romaine when a combined hot-air and hydrogen balloon exploded. They were the first aeronauts to die on active service.

Within three months of the first Montgolfier flight at Annonay, Jacques Charles, a 36-year-old teacher at the Sorbonne, built and flew a balloon filled with hydrogen (which Henry Cavendish had isolated in 1766 and Charles could make more expertly than his rivals). Among the crowd watching this unmanned take-off was Benjamin Franklin, the United States ambassador, who answered a scornful question *What's the use of a balloon* with the historic retort *What's the use of a new-born baby*? Nevertheless the balloon, which smelled hellishly of sulphur, was ''lynched'' by the villagers where it landed. Ten days after de Rozier's ascent in Paris Charles personally went up with Nicholas Robert, one of two brothers who had made the fabric, in a basket slung from an envelope of rubber-impregnated silk controlled by far more sophisticated equipment than the Montgolfiers had devised. With a battery of instruments, Charles began observations which were to result in his own professorship and the enunciation of Charles' law relating the volume of a gas under constant pressure to its absolute temperature (the scale placing absolute zero at −273°C).

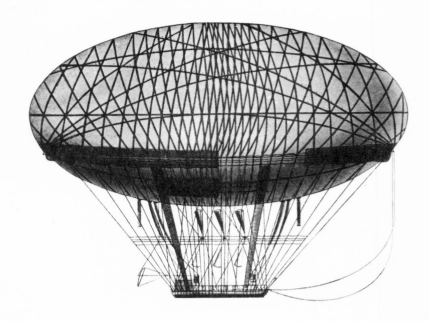

The year 1783 was a remarkable one for aeronautics in France. Near the end of the year, a paper on *The Equilibrium of Air Machines* was submitted by Lieutenant Jean-Baptiste Marie Meusnier of the Corps of Military Engineers, 29 years old and an advisor of Professor Charles and the Robert brothers. The *Académie des Sciences* thought so highly of Meusnier's paper that they commissioned him to continue his theoretical studies. In 1785 he produced a striking summary of the essentials of a lighter-than-air dirigible. He declared that the balloon should not be round, but long and slender. Since the airship retained its shape only by the pressure of the gas inside it, when irreplaceable gas was released to control height the craft was liable to lose its shape and go out of control; therefore, Meusnier proposed a double envelope with the outer chamber filled with slightly compressed air – the ''ballonet'' principle later adopted in non-rigid airships. He had used this device in practice on an elongated balloon which Charles and the Roberts had made and launched on July 15, 1784, and on a second cylindrical Robert balloon which flew from the Tuileries to Béthune in six-and-three-quarter hours on September 19. Meusnier's definitive design of 1785, meticulously drawn, postulated an efficient power unit to achieve steerage way. He provided three propellers, but, in the temporary absence of a suitable engine, he called for the propellers to be operated manually by 80 men, for whom he provided accommodation. Apart from the insuperable traction problem, Meusnier's fine design remained an inspiration to all future airship builders.

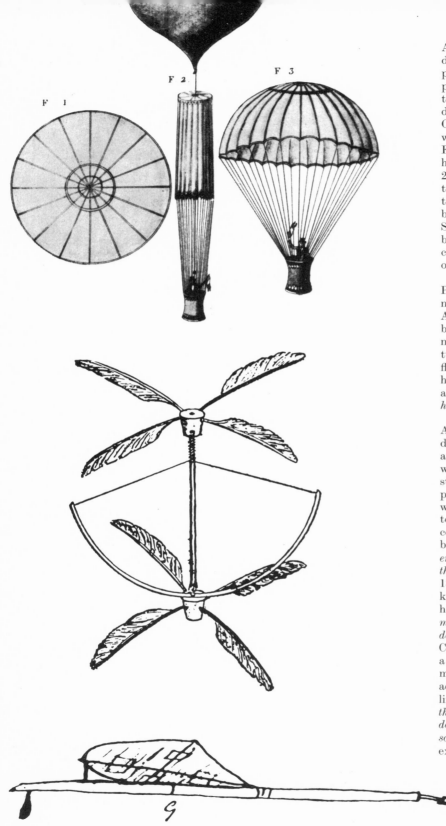

A revival of interest in parachuting accompanied the astonishingly swift development of the balloon in France in 1783. Sébastian Lenormand, a physician at Montpellier, first jumped from a treetop using only two parasols, but in December, 1783, he used a much more sophisticated technique to build a saddle hanging from a conical parachute 14 feet in diameter, with which he dropped safely from the tower of Montpellier Observatory. In the pursuit of ballooning, a number of aeronauts escaped when suitably shaped remnants of burst balloons wafted them to safety. Finally an experienced balloonist, Andre-Jacques Garnerin, constructed his own parachute and tested it by jumping from a balloon. On October 22, 1797, Garnerin rose over the present Parc de Monceau in Paris, in a tub attached, beyond its parachute, to a hydrogen balloon. After cutting the lines at 1,000 meters, he descended safely. Garnerin supported himself by further demonstrations, such as his first drop over London on September 21, 1802. His parachute was comparable to the modern design but had no vent in the crown to release the compressed pillar of air, and consequently induced sickening and sometimes dangerous pendulum oscillations.

Paris consolidated its lead in aeronautics when on April 23, 1784, the naturalist Launoy and his mechanic Bienvenu demonstrated to the Académie a practical model of a helicopter with two rotating helical blades powered by a bow spring. Leonardo da Vinci had designed – but never built – a hand-held toy helicopter working by clockwork; and only two years before Launoy, Jean-Pierre Blanchard had built – but never flew – a manually operated man-size helicopter. The Launoy/Bienvenu helicopter was to be virtually duplicated in 1796 by Sir George Cayley, a 22-year-old Yorkshire baronet, who later said it was *the first idea I ever had on the subject of mechanical flight.*

All through the 17th century the main use of kites was for aerial firework displays. By the beginning of the 19th century a kite was being used as an airplane, in the strict historical use of that word as meaning a fixed wing in an aircraft. Halfway between these dates, Isaac Newton, while still a schoolboy, had played pranks with kites as fire carriers but had pondered over their use as aerofoils. The same candlelit paper lanterns which he had carried to school on dark winter mornings became affixed to the tails of the kites he was expert at flying. With them, he frightened country folk who believed they were ominous comets. But, said his biographer, *Even when he was occupied with his paper kites, he was endeavouring to find out the proper form of a body which would experience the least resistance when moving in a fluid.* Well over a century later, in 1804, Sir George Cayley, the "Father of Aerial Navigation," sketched a kite-wing glider which the aeronautical historian Charles Gibbs-Smith has called *the first modern configuration aeroplane of history, with fixed mainplane, and combined and adjustable rear rudder and elevator; the description of its flight is also the first in history of true aeroplane flight.* Cayley said he used *a common paper kite containing 154 sqr. inches* with a cruciform tail of 20 square inches set at an angle of six degrees to a mainspar rod, with a movable forward weight which provided an adjustable center of gravity. When he had learned to fly it in a straight line he crowed: *It was very pretty to see it sail down a steep hill, and it gave the idea that a larger instrument would be a better and a safer conveyance down the Alps than even the sure-footed mule, let him meditate his track ever so intensely.* Cayley used the kite as a wing for many future models, experimenting with dihedral setting and extremity wing panels.

Rockets had a functional revival in the 18th century. Even before the manufacture of cannon, they had been used as military projectiles in the 12th and 13th centuries, serving as incendiaries rather than as explosives. Later, they diminished in importance and were developed almost solely as fireworks. Enthusiasts, principally German artillery officers, made some refinements over the centuries: fins for stabilizing their flight in 1590, an increase in possible size up to 132 pounds with a 16-pound gunpowder charge in 1668. Yet it was an Indian adventurer, Hyder Ali, Sultan of Mysore, who in the 1770s developed war rockets to a new potential. He used hammered iron cylinders to contain the gunpowder, and with increased thrust achieved a range of three quarters of a mile. The rockets were fired in concentrated salvoes and were most effective against cavalry. Hyder's son Tipu further developed rocketry and quadrupled his French-trained rocket corps to a strength of 5,000 men. The illustration shows a rocket bearer in this corps in 1798. Rockets were used by Tipu to cause considerable damage among the British forces besieging his capital, Seringapatam, in 1792 and 1799. These military incursions prompted British inventors to take up the development of the rocket.

William Congreve was son of the Comptroller of the Laboratory of the Royal Arsenal at Woolwich, and eventually succeeded his father in that post. He was 27 years old in 1799 when news of the destruction caused by Tipu Sultan's rockets at Seringapatam reached the British Army at home. He studied the military use of rocketry and developed a rocket capable of carrying a six-pound warhead 2,000 yards. In 1806 Congreve conducted a successful raid on Boulogne with 200 rockets discharged from ships fitted to launch them. A year later, Congreve rockets were directed against Denmark, an ally of Napoleonic France; half the city of Copenhagen was burnt to the ground after a bombardment of 25,000 rockets, and the British captured the Danish fleet. In 1813 a specialist Rocket Brigade had been set up within the Royal Regiment of Artillery and was highly effective in the victory over Napoleon at Leipzig. The following year, "the rockets' red glare" was applied against untrained American troops and led to the breakthrough that culminated in the burning of Washington. By this time Congreve's rockets, both incendiary and explosive, ranged from light projectiles which scattered 48 carbine balls to heavier rockets carrying 18-pound shells or 12-pound spherical bombs. Their ideal use was in salvoes of 50 fired every 30 seconds.

The first striking and practical early application of screw propulsion was in a submarine. Navigation *under* water, being slow and arduous, was at first reserved for clandestine purposes. Apart from rare diving for sunken treasure, the submarine was mainly a war vessel. It was pioneered by a man of peculiar enthusiasms. David C. Bushnell was born in Saybrook, Connecticut. Since he had no interest or skill in agriculture, he sold the family farm after his father died in 1771, and at age 29 used the money to go to Yale. His principal exploit there was to demonstrate that gunpowder could be exploded under water. By 1775 he had built a submersible craft, an upright oak sarcophagus shaped like a top. Because it also resembled two turtle shells fastened together tail downward, it was called Bushnell's Turtle. It had two screw propellers, one horizontal and one vertical, which the one-man crew worked by hand. A foot valve admitted water for submersion and two hand pumps ejected it. Bushnell's only projected use for his Turtle was to attach primitive limpet mines to the underwater hulls of enemy ships, and in 1776 the enemy was Great Britain.

The mine was a wooden magazine of gunpowder with a clock mechanism for ignition. It was attached to the outside of the submarine and connected to a wooden screw, turned from inside the submarine, until it was driven into the enemy ship's bottom. The entire operation had to be completed in half an hour for the crewman to subsist on the air left in the vessel after submersion. Bushnell himself was not sturdy enough to man the submarine, but a Sergeant Lee attempted on three occasions in 1776 and 1777 to blow up British ships in Boston Harbor, off Governor's Island, New York, and in the Delaware river above Philadelphia. Each time he failed, mainly because he was not very good at planting the mines on the hulls. Bushnell, still an explosives enthusiast, finished his military career commanding the Corps of Engineers at West Point. In addition to making submarine navigation an actuality, David Bushnell had demonstrated the mechanical transfer of energy into motion in a craft. He did this by operating a hand-turned screw propeller, and soon Blanchard was to apply handscrewing to a propeller in a balloon.

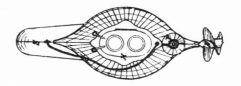

While Bushnell was at work on the Delaware, a 25-year-old infantry captain, Claude Francois Dorothée, Marquis de Jouffroy d'Abbans, was applying steam to marine transport on the river Doubs at Besançon, the fortified town in eastern France with a not-irrelevant watchmaking tradition. In 1776, Jouffroy fitted a single-acting Watt-type steam engine into a 42-foot boat, and was able to report progress, though inconsistent, against the current on that fast-flowing river. On July 15, 1783, Jouffroy made a successful trip up the Saône from Lyon in a practical steamboat. His horizontal double-acting steam engine had been developed independently of Watt, who produced his model in the same year. It rotated two paddle wheels with its 25-inch-diameter cylinder inside the boiler, to drive the 140-foot craft *Pyroscaphe*. Denis Papin had put an atmospheric engine into a paddle boat in 1707, and saw it destroyed by German watermen. But the *Pyroscaphe* was a practical success and worked on the Saône for 15 months. However, in marked contrast to their extraordinary enthusiasm for balloon travel, the French *Académie des Sciences* gave Jouffroy no active encouragement, and he abandoned steamboats for many years.

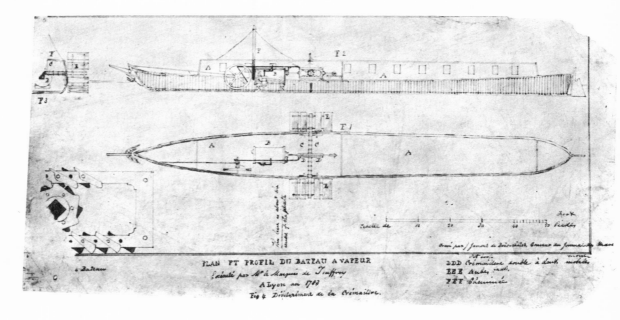

FITCH'S STEAMBOAT.

Mechanical marine propulsion had existed since the adoption of the oar as a lever, and James Watt had spoken of the screw propeller as the *spiral oar*. Rotating paddles were merely a reversal of the principle of the water wheel. But an early steamboat experimented with propulsion by driving, lifting, and dipping vertical paddles, perhaps styled after the war canoes of the neighboring Indians. This was the 30-foot craft of John Fitch tested in 1786 and publicly demonstrated on the Delaware River at Philadelphia on August 22, 1787, to delegates to the Constitutional Convention. The steam engine, made by Johann Voight, a Philadelphia clockmaker, managed to propel the oars with enough strength to gain a slight advantage over wind and tide. Fitch and Voight redesigned the craft the next year, finally opting for a set of three duck-feet paddles placed at the stern, and driven by a beam engine with a single cylinder of 18 inches diameter, giving the boat a speed of seven miles an hour. On July 26, 1790, Fitch began a regular thrice-weekly passenger service from Philadelphia to Trenton in a vessel of this design which he named *Experiment*. Fitch failed to get financial backing for commercial ventures with steamboats and in 1798, after rejection of a scheme to promote screw-propelled craft, he committed suicide.

The ratchet-wheels system used by Jouffroy to provide rotary motion for the *Pyroscaphe* was also employed in a road locomotive demonstrated in Edinburgh in 1786 by William Symington, a 25-year-old engineer. Symington's steam engine caught the fancy of Patrick Miller, a wealthy banker. Miller had already been working on manually driven rotating paddles for boats. He engaged Symington to adapt the engine of his road locomotive for marine transport. The young man installed an engine in a 25-foot twin-hulled catamaran-type pleasure boat. The weight of the boiler was taken on one hull, and on the other was a two-cylinder atmospheric engine using Watt's condenser. Chains driven from pawl-and-ratchet gear rotated two paddles placed in tandem in the space between the hulls. On October 14, 1788, this craft successfully steamed on Dalswinton Loch in Dumfriesshire. Alexander Nasmyth, the painter and father of the engineer James Nasmyth, was one of the six passengers aboard. Another was the poet Robert Burns. Miller commissioned a larger engine which was fitted in 1789 in a 60-foot boat which operated successfully as a tug on the Forth and Clyde Canal, where 13 years later Symington's *Charlotte Dundas* worked as a tug with the first direct-acting steam engine, promptly adapted by Robert Fulton.

Bliss was it in that dawn to be alive. But to be young was very heaven! wrote Wordsworth on the beginning of the French Revolution. For his contemporary Claude Chappe, the bliss was mixed. Chappe had a vocation for the Church, but he was ardent for reform. After the Revolution he left his seminary and applied himself to long-distance communication on a strictly terrestrial basis. He first experimented with an electric telegraph. Some 40 years previously, in open country, another French cleric, the Abbé Nollet, had arranged a contingent of Carthusian monks in a circle more than a mile in circumference, with each monk connected electrically by iron wires. The Abbé discharged electricity into the circuit, and was intrigued to see all the monks jump simultaneously with the shock.

Chappe proposed that impulses encoding messages should be transmitted, not through a circle of monks but down a straight line of connected telegraph stations. He found it speedier and more practical to concentrate on visible signals relayed from point to point in a chain of towers; each station was some 10 miles apart, manned by operators equipped with telescopes, who passed semaphore signals conveying coded messages. In 1793 the French Revolutionary Government had appointed Chappe as Ingénieur-Télégraphe and commissioned him to set up stations between Paris and Lille, a distance of some 160 miles. The first important message, passed in August, 1794, informed Paris of the recapture of Le Quesnoy from the Austrians. It arrived in 20 minutes: a speed of 480 miles per hour. Eventually France had a network of 3,000 miles of semaphore stations, and there were minor installations on telegraph hills in Great Britain and the United States. But Chappe, who had seen his first telegraph transmitters senselessly destroyed by the Paris mob in fury against the army, did not supervise much of the construction. Overwhelmed with business worries, he killed himself in his workshop in 1805 at the age of 41, while Wordsworth, who was to survive another 45 years, was writing *Intimations of Immortality*.

In England the law of patents was long governed by the unpredictable whims of the Monarch and his disposition to grant a "Privilege". But in the second half of the 18th century an inventor acquired a legal right to exploit his patented innovation. In the 25 years after 1760 more patents were issued in England than during the preceding 150 years. Many of them fell by the wayside, unexploited, to be picked up later or to be published by others at a more advantageous time. A casualty of 1790 was the sewing machine patented by Thomas Saint of London. Intended by Saint to be used for leather-working, and to be driven alternatively by a hand crank, a water wheel, or a steam engine, it was never developed, nor subsequently noticed until 84 years later. By that time Isaac Singer, with only three years to run on his own patent and others he had borrowed, had cleaned up the market.

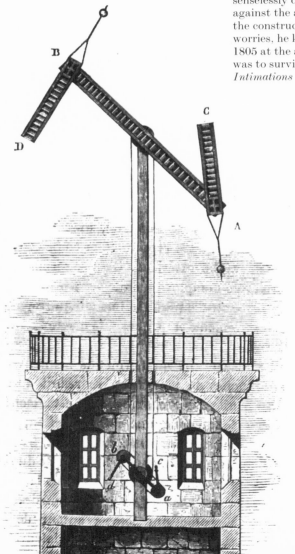

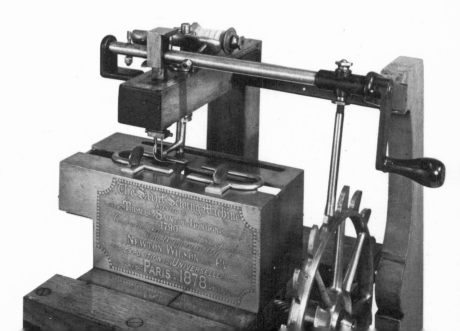

In the newly independent United States, entrepreneurs who seized on new patents were as obstreperous and litigious about paying royalties as Hargreaves and Arkwright had found them in England. They were sufficiently crafty in avoiding payment of the fees due to Eli Whitney for his cotton gin that he was forced, in his poverty, to a far more revolutionary innovation. In doing so, he changed the pattern of manufacture, which led to the mass production of interchangeable parts known throughout the 19th century as the *American system*. Whitney, who was a Massachusetts farmer's son always clever with his hands, left Yale at the age of 27 in 1792 and went south to seek work as a tutor. His hostess in Georgia mentioned the one handy gadget every cotton planter dreamed of – an apparatus to extract the seeds which remained entangled in the short fibers of greenseed upland cotton. Within 10 days Whitney had worked out a successful design, and by the spring of 1793 he had built it. It consisted of a spiked cylinder revolving under a slotted cover, over which the raw cotton was passed. The revolving teeth caught the fibers of the cotton and drew them through the slots, which were too narrow to let the seeds pass, so that they were left outside. A roller fitted with brushes took the fibers off the toothed roller and an air fan sent them on for the next process.

This brilliantly simple machine was known as Whitney's cotton gin (*gin*, meaning a mechanical contrivance, was used in English long before Chaucer). It was patented in 1794 and was taken up with gusto in the South. The speed with which it cleaned cotton produced not only a fiftyfold increase in output but made economically profitable the planting of short-staple cotton – the only type that would grow far from the sea – in upland areas, which previously had been impractical. The speed of the process encouraged the planting of all types of cotton, particularly the high-quality long-staple sea-island cotton which America had not previously produced. Within 16 years of the introduction of Whitney's cotton gin the annual output of raw cotton in the United States rose from two million to 85 million pounds. The planters were profiting hugely. But they pirated Whitney's idea and paid him no royalty. In dire financial embarrassment, Whitney proposed to, manufacture 10,000 small arms for the U.S. Government, using a precision system of interchangeable parts to be assembled by unskilled labor. A contract was signed in 1798 and Whitney managed to get capital advanced on the strength of it. He was overly optimistic: he needed more skilled workers than he had anticipated, and he had to adapt existing machine tools to make parts to an unprecedented standard of precision. The contract was fulfilled, though tardily. A second contract was agreed upon, and soon the U.S. Government had introduced Whitney's system into all its armories.

Just as Whitney's cotton gin was never fundamentally improved from its inception, the contemporary introduction of Jacquard's loom has never, after its initial amendments, been changed in essentials. Joseph Marie Jacquard had grown up in Lyon as an artisan concerned with type-founding and cutlery. But he was bequeathed a property which gave him an interest in weaving and he began to experiment with looms. By the time he was in his late 30s, he had lost his inheritance, and the French Revolution was at the gates of Lyon. The times were not conducive to the technical improvement of weaving. Not until 1801, when he was nearly 50, did he produce the original Jacquard loom for pattern weaving. Though it had to be perfected and modified over the next few years, it was speedily adopted – 10,000 looms were sold within the first 10 years – and to this day it remains the only machine on which complicated patterns can be woven.

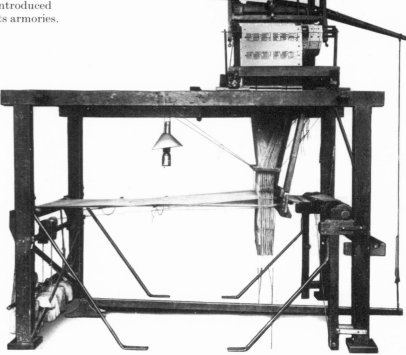

The growth of industry was matched by a growth in the clerical administration it demanded. The laborious practice of making handwritten letters and accounts was made even more irksome by the need to keep a further set of handwritten copies, as records of the various dealings. James Watt, and his partner Boulton, found that as their business expanded, so did the problem of keeping copies. Watt devised a method of copying by machine, using his knowledge of chemistry to produce a formula for ink and unsized paper which, by pressing one sheet upon the original, gave a reverse impression which could be read by holding the paper to the light and looking through from the other side (the paper being very thin). In 1780 Watt produced and patented two types of press, one a roller, the other operated by a screw. The roller copying press and special ink was the first to be manufactured by a separate company, James Watt & Co., and a hundred and fifty were sold to bankers, businessmen and merchants in the first year of manufacture. The screw model was then developed, and the two versions became an indispensable standard item of office equipment until the advent of the typewriter and carbon paper, at the end of the 19th century. Unwittingly Watt had produced an offset printing machine. In 1798 Alois Senefelder discovered the principles of "lithography", i.e. printing from a grease-attracting image drawn on stone. Not until 1875 did Robert Barclay patent his offset-lithography process, for printing from stone to rubber rollers and then back onto tinplate.

The fact that paper could be manufactured from the fibers of trees and plants, established by Jacob Schaffer, a German Pastor, led Nicholas-Louis Robert to devise a machine for its manufacture, which turned the wood pulp, or fiber, into strips of paper up to 12 feet wide and 50 feet long, using a continuous copper or brass wire belt with a series of rollers to squeeze the pulp, and an oscillating motion to remove the water. Robert sold his 1799 patent to St. Leger Didot, whose brother-in-law persuaded the English papermakers H. & S.Foudrinier to build the first machine. The textile industry had for some time been printing on rolls of calico and other materials, using mechanized cylinder presses (invented by Thomas Bell and first installed in a Preston mill in 1784–5), and printing up to five colors. The use of printing ink upon the continuous roll of paper produced by Robert's machine was first applied, for publicity purposes, by J. & T.Gilpin in Delaware in 1817; the first American manufacturers of "endless paper" and the first to feed a reel of it to a rotary printing press.

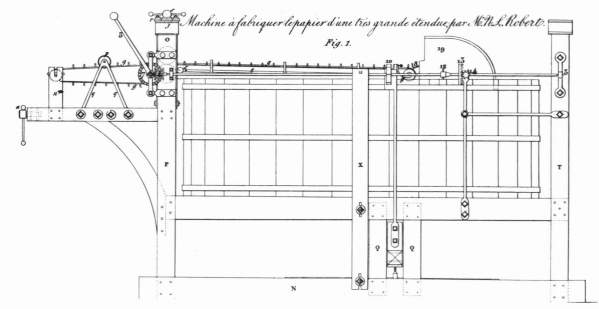

Machine à fabriquer le papier d'une très grande étendue, par Mr. N. L. Robert.

Fig. 1.

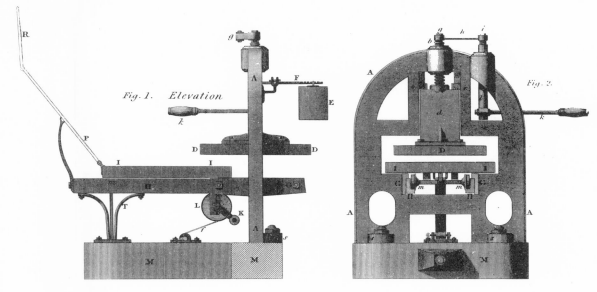

The STANHOPE or IRON PRESS.

Fig. 1. Elevation

Fig. 2.

Handprinting was advanced by the improvement in metal-casting techniques which made possible the adoption of cast iron in place of wood for the construction of the press, and by the use of a combined screw and lever movement. The greatly increased pressure made possible thereby allowed the bed of the machine to be increased in size so that it could print twice the number of pages in one operation. These improvements were the work of Charles, 3rd Earl Stanhope, whose other interests included politics, philosophy, and science and who earlier had designed and produced two calculating machines.

His enthusiasm for the printing press was not matched by his practical engineering skill, and he employed Robert Walker to assist him in developing the Stanhope Iron Press, late in the 18th century; work on the prototype was completed by 1804. He did not patent the design, as he wished the printing trade to use it freely, and once some minor troubles had been overcome, the London *Times* was printed with what it described as a *battalion* of Stanhope presses.

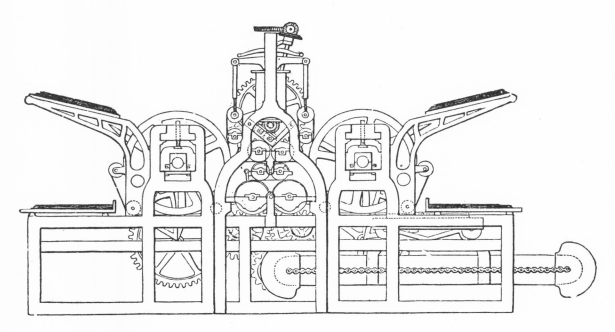

Daily newspaper production requires speed for only a few hours each day. The London *Times*, experiencing a tremendous increase in circulation, sought to mechanize its presses and secretly installed the first steam-driven double machine at Printing House Square in 1814, going into printing production on November 29. This machine was the work of Friedrich Koenig, a printer and bookseller from Thuringia, who had first produced a power-driven press in Suhl, Saxony in 1803, in order to improve upon the traditional methods of inking. Moving to London, Koenig entered partnership with Thomas Bensley, a well known printer, and two others, and was joined by Andreas Bauer, a fellowcountryman and skilled engineer. Their first machine was a failure due to retention of the traditional platen principle, and Koenig next tried using the impression cylinder. Two such machines were ordered for Printing House Square, with two cylinders and one type forme, so that output was doubled by printing on both trips of the reciprocating carriage. *The Times* was printed at the rate of 1,100 impressions an hour by Koenig's first machine, and the hand-operated Stanhope presses were dispensed with.

For 5,000 years the output of a normal source of light was one candlepower. In 1782 the Swiss physicist Aimé Argand produced a lamp-wick burner which multiplied that basic illumination by 12. Air was led through the cylindrical wick of an oil lamp, and rose in a steady current on the outside of the wick, controlled by a glass chimney. Highly efficient combustion of the oil was achieved with a smokeless flame. Domestically, the Argand lamp, which originally burned fish oil, used so much fuel that it was a luxury. But it revolutionized the illumination of marine lighthouses. Before Argand most mariners consumed up to 300 tons of coal a year by burning fires at the top of their lighthouses, whereas Smeaton's Eddystone lighthouse burned 24 tallow candles at a time. The virtue of Argand's burner was that its flame was smokeless, bright, and extendable. Soon lamps with 10 concentric wicks were designed. With its fish-oil fuel yielding first to vegetable oil, then to mineral oil (paraffin), the Argand light continued in constant use around the coasts of the world for over a century.

At first combustible gas had been distilled from wood. In 1801 the 34-year-old French engineer Philippe Lebon, who had spent some of his youth among charcoal burners, used wood gas to fuel his thermolamps, with which he illuminated the Hôtel Seignelay in Paris. But coal gas was technically easier to distil. A pioneer in its use was William Murdock, the Cornish steam-engine builder. Murdock had used coal gas to light his house in Redruth from 1792. In 1801, simultaneously with Lebon and probably to celebrate the same event – the brief peace between Britain and France marked by the Treaty of Amiens – he used coal gas to illuminate the engine house of the Boulton & Watt factory in Soho, Birmingham. With a younger contemporary, Samuel Clegg, Murdock was mainly concerned with exploiting gas illumination (utilizing the Argand burner) to keep factories running efficiently during the night, and in 1806 he lit a textile factory in Salford. Street lighting by gas was introduced in London in 1807, when Pall Mall was lit by a commercial company which eventually became the Gas Light and Coke Company.

In 1791 Luigi Galvani, a 54-year-old theologian *manqué* and for 25 years professor of anatomy at Bologna, published a paper setting out the theory that a form of electricity existed in animal muscle. Galvani had come to this conclusion after long years of experiments which had started when the leg muscle of a dead frog twitched under his scalpel as he was dissecting it, while an electrostatic machine sparked in the vicinity. (At that time the only known electrical generation was caused by friction between such materials as silk and amber, and this static electricity could be stored in Leyden jars.) Galvani went on to expose animal muscle to various controlled conditions, including thunderstorms. Awaiting electrical storms, he would hang frogs' legs from his balcony on brass hooks which gave the flesh occasional contact with an iron lattice. He noticed that the muscles twitched whenever they connected the two metals. Galvani had isolated a new electricity, distinct from static electricity, capable of flowing in a current. He called it animal electricity, believing that its source was animal muscle. He died in 1798, a largely discredited man, disbarred from his university chair because he would not swear allegiance to the new Cisalpine Republic instituted by Napoleon Bonaparte.

Argands.

One of Galvani's strongest theoretical opponents had been Alessandro Volta, professor of physics at Pavia, who was made a count by Napoleon in return for his loyalty. To discredit the theory that electricity was generated by muscle when the animal tissue touched two different metals, Volta used metal plates *not* connected by muscle. He detected an electric current although no obvious conductor existed, save air. He defined this current as metallic electricity. Volta had already perfected a device, improving on the Leyden jar, for accumulating and storing charges of static electricity. In 1800 Volta demonstrated a battery which would generate a current of the new electricity, using pairs of connected zinc and silver discs immersed in, and chain-linking, bowls of weak acid. A chemical reaction had produced an electric current. Volta was a fellow of the Royal Society, and he wrote to London to inform the president of his achievement. Within weeks, William Nicholson, an English chemist, built his own Voltaic pile and, by passing a current through a bowl of water, achieved the electrolysis of water, releasing its elements while hydrogen and oxygen bubbled off. An electric current had produced a chemical reaction.

Soon Humphry Davy, the young, flamboyant, and popular professor-lecturer at the Royal Institution in London, built a Voltaic pile of 2000 plates, with which he electrolyzed a solution of caustic potash and isolated the metal potassium. Later he isolated sodium. In 1809 Davy went on to apply the new electric current to produce light. He used two carbon rods as electrodes, let them briefly touch, and then drew them slowly apart. Electricity continued to flow from one rod to the other in a brightly glowing arc. Davy, in lecture-demonstrations he gave to swooning audiences, advanced the arc in 1810 to cover a distance of some three inches, while the carbon steadily burned away. This discharge, in plain air and without the shelter of any lantern, was the first arc lamp. It was impracticable, though street lighting by arc lamp was a commonplace well into the 20th century. For most of the 19th century extensive public illumination was dependent on coal gas. Davy went on to invent that other "miners' friend," the safety lamp, in 1817.

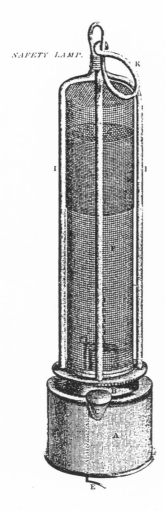

SAFETY LAMP.

About 1800, a canal was proposed to link the industrial Wandle valley, south of London, with the Thames at Wandsworth. This resulted in the construction of the Surrey Iron Railway – the first public railway. The inadequate roads of the area could not cope with the flow of raw materials needed by the various factories and mills, and local business-men approached the canal engineer William Jessop to survey the route. He reported that a canal would deplete the water resources of the Wandle River (from which it would be fed), preventing them from using the Wandle to power their mills and factories. As an alternative he proposed an Iron Railway, and this was agreed upon, with horse haulage and wagons with flangeless wheels running upon flanged cast-iron rails. A Parliamentary Bill, setting up the Surrey Iron Railway Company, received the Royal Assent in May 1801 and by July 26, 1803, the whole line from the Wandsworth dock to Croydon was opened. It was public, in the sense that its Act of Parliament stipulated that *All persons whomsoever shall have free Liberty . . . to pass upon and use the said railway with Waggons or other Carriages* on payment, and subject to the Company's rules and regulations. Jessop suggested it be extended to the South Coast (at Portsmouth) but the only extensions actually built were a branch to Hackbridge and the Croydon, Merstham and Godstone Iron Railway, which reached Merstham Quarries in 1805. The line never carried a steam locomotive, and did not come up to the commercial hopes of its promoters; it was closed on August 31, 1846, as *the traffic along the said line has ever since the completion thereof been very small, and has of late years been diminishing.* The railway had "up" and "down" lines so that it could operate continuous traffic in both directions. This was to become an important feature of railway operations in the future.

The Cornish engineer, Richard Trevithick, invented and patented a high-pressure stationary steam engine which was small enough for him to envisage placing it upon wheels. He made three models in 1797, one of which had four wheels, and was sufficiently encouraged to build a road locomotive, which ran on Christmas Eve, 1801, carrying a party of seven or eight enthusiasts, who hung on as it steamed up a steep hill. A steam road carriage followed, and was demonstrated in London at speeds up to 10 mph. An invitation to build a tramway locomotive for the Pen-y-Darran tramway in South Wales was accepted and he devised a crude "tram-waggon" with one of his high-pressure steam engines (actually constructed to power a hammer) mounted upon it. A wager of 500 guineas as to its success in hauling a load was an added incentive. On February 20, 1804, Trevithick wrote "To Mr. Giddy":
Sir, – The tram-waggon has been at work several times. It works exceedingly well, and is much more manageable than horses. We have not tried to draw more than 10 tons at a time, but I doubt not we could draw 40 tons at a time very well . . . The engine, with water included, is about 5 tons. It runs up the tramroad of two inches in a yard forty strokes per minute with the empty waggons. The engine moves forward 9 feet at every stroke . . . The steam that is discharged from the engine is turned up the chimney about 3 feet above the fire, and when the engine works at forty strokes per minute, 4½ feet stroke, 8 inches diameter of cylinder, not the smallest particle of steam appears out of the top of the chimney.

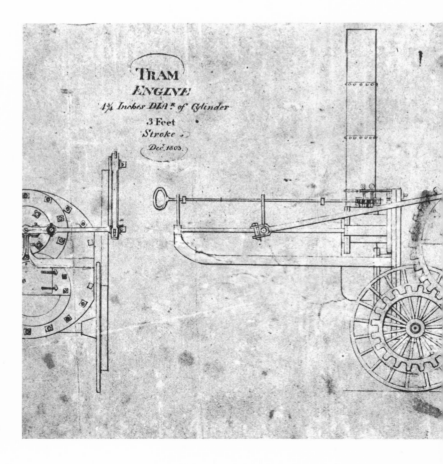

Trevithick won his bet on February 21, 1804, when the locomotive ran a course of less than 10 miles downhill in a very leisurely four hours. For a short time the locomotive continued to run, with up to 40 tons, but it proved too heavy for the tram plates and was soon relegated to a stationary existence. A second Trevithick locomotive, with flanged wheels, was built for Wylam Colliery, near Newcastle, but never put to regular work as again it proved too heavy. A third was built by John Urpeth Rastrick, of Stourbridge, in 1808, to Trevithick's design. Trevithick, then in London, enclosed a space of ground on the site of Torrington Square, within which he laid a circular railway. He opened this to the public, who were invited to travel in an open carriage behind the locomotive for a fee of one shilling. It ran for some weeks until the breakage of rail caused him to abandon it. The locomotive had the nickname of "Catch me who can," and is said to have reached 12 mph. It had a single vertical cylinder driving directly upon both rear wheels by return connecting rods, and weighed eight tons with its cast-iron boiler. Problems of excess weight and the inability of existing tracks to withstand the strain seem to have caused Trevithick to turn his attention away from further developments. But the seed was sown, for one of those who saw his engines in action was the young enginewright George Stephenson.

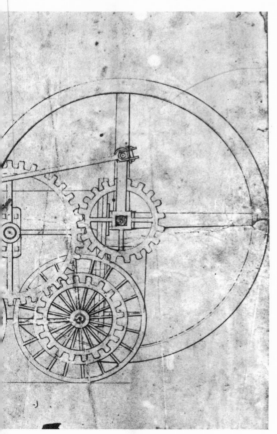

If Joseph Bramah had done nothing beyond improving the recently patented U-valve, forming a stinktrap below the pan of the water closet and introducing simultaneous flushing, he would rank as a public benefactor. These developments kept his firm in business for over a century after the registration of his patent in 1775. But he did much more, particularly in precision engineering and hydrostatics. He was a Yorkshire farmer's son who, crippled by an accident, was apprenticed to a London cabinet-maker and later went into business on his own. A small part of his trade was to fit the houses of the gentry with water closets (a term unknown when he was born in 1748), and his always agile mind was responsible for the comparative perfection of his sanitary system. With his interest in hydrostatics, he created products as wide-ranging as a primitive fountain pen, the suction beer pump, and the simple but sophisticated hydraulic press.

One hundred and fifty years earlier, Blaise Pascal had observed that a small-bore piston pushing far down into a cylinder would raise a large-bore piston in its separate cylinder if both were in a reservoir containing a non-compressible liquid. Bramah recognized this as the function of a lever, and devised the hydraulic press, the ancestor of many mechanical devices. It was first used, on Bramah's suggestion, to extrude continuous lead piping. Bramah gave engineers the power to convert the intermittent action of a small-bore pump into a steady and virtually unlimited pressure. But he first had to overcome the immense difficulty of making a leak-proof cylinder surrounding the ram by using a self-tightening collar, the leather cup washer. For this and many other refinements he was indebted to his assistant, Henry Maudslay.

Maudslay was 24 when this patent was specified in 1795. He had begun to earn his living at the age of 12 by filling cartridges in Woolwich Arsenal. Within six years he was an accomplished smith, and Bramah took him on as an aide in constructing precision machinery for the manufacture of a famous tumbler lock which, when perfected, was never picked for 67 years. Because of the machining accuracy necessary for the construction of the sliding parts of this lock, Bramah and Maudslay began their lifelong task of building precision machine tools. Previously, metal-cutting tools had been held by hand against the turning workpiece on the pattern of the woodworking lathe. An inadequate hand rest was later installed. Bramah set out to build machines that would replace manual control. With Maudslay he produced in 1794 the lathe slide rest capable of holding the tool in a fixed position. Its immediate application was to permit the controlled advance of a cutting tool which effected mechanical regularity in cutting a screw on a large metal bar or cylinder. In 1798 Maudslay set up on his own. By 1800 he had advanced precision engineering with his increasingly accurate screw-cutting lathe, which opened the way to the manufacture of identical, thus interchangeable, metal parts, and led to mass production. He went on to build the first metalworking lathes with automatic movement of the carriages on precision-controlled slideways, and added a number of striking developments in the construction of accurate machine tools.

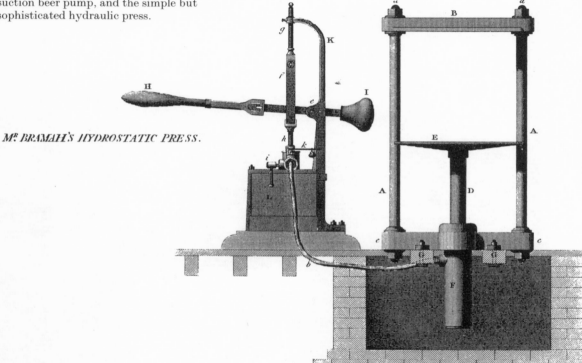

Mr Bramah's Hydrostatic Press.

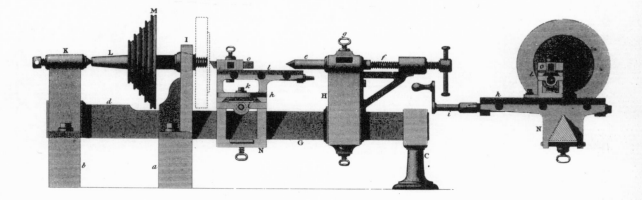

Maudslay's first screw-cutting lathe had gained the attention of Marc Isambard Brunel, a former French naval officer who had left the service after the first excesses of the French Revolution and set up a new career as an engineer in New York. After building the city arsenal, he had emigrated to England at the turn of the century; there, he persuaded the Admiralty to adopt his scheme for the mechanical manufacture of pulley blocks for ships' tackle. With 44 Maudslay machines performing different operations, all powered by one 32 horsepower steam engine, he revolutionized the Portsmouth dockyard. He did so by making 130,000 pulley blocks a year with a labor force of 10 machine operators, as against the 110 skilled craftsmen previously employed. This exercise was the first example of practical mass-production methods. The system was maintained at Portsmouth for 130 years. It served the fleets commanded by Nelson and by Jellicoe, a remarkable continuity between the battles of Trafalgar and Jutland. Maudslay continued to refine the process of metal machining. He produced a screw micrometer with an accuracy of 1/10,000th of an inch. He raised the standard of British engineering by transmitting his insistence on accuracy to his workmen and his professional trainees. Among those who served with him were Richard Roberts, Joseph Whitworth, and James Nasmyth. Nasmyth, who did not set up on his own until after Maudslay's death in 1831, paid posthumous tribute to a master *whose useful life was enthusiastically devoted to the grand object of improving our means of producing perfect workmanship and machinery.*

The Portsmouth project had shown that machining wood was a simpler proposition than machining metal. Repetitive woodworking was further advanced in 1818, when Thomas Blanchard, a machinist in Worcester, Massachusetts, built a copying lathe for turning gun stocks. Until then, gun stocks had been a desirable timber product for mass production but had not been tackled because of their irregular shape. The lathe used the pantograph principle, having a rotary cutter governed by a tracing wheel fixed against the master pattern. This copying lathe, soon installed at Springfield Armory, had a life of 50 years and saw the disunited States through the Civil War. But long before that the duplicating principle had been applied to working metal. Metal is mainly shaped by rotary motion or reciprocal action. The manual movements which reciprocal-action machines most closely imitate are sawing, filing, and planing. Joseph Bramah patented a wood-planing machine in 1802 and later, when the demand for his burglar-proof locks made speedy manufacture essential, he introduced a metal planer. A planer which still exists in the Science Museum in London was launched in 1817 by Richard Roberts. Roberts was a tough, individualistic artist who introduced more mechanical innovations than anyone else in the 19th century, but whose influence as a technological leader was short-lived. In the prime of his life he inspired others, in the sense in which Bernard Shaw's Serpent in Eden declared to Eve: *You see things; and you say "Why?" But I dream things that never were, and I say "Why not?"*

Roberts first worked in Staffordshire as a patternmaker for the ironmaster John Wilkinson. Wilkinson had produced the first precision boring machine, and cannon he had bored were used by Wellington and by the French during the Napoleonic wars. Ironically, Roberts had deserted the militia and joined Maudslay in London to escape the war's carnage. In 1816 he left Maudslay and cannily set up on his own in Manchester where the textile makers were the best paymasters. During the next year, when he built and used his metal planer, he also constructed an historic screw-cutting lathe. His many additions to the machining and machine-tool equipment of the textile industry included a self-acting mule which he devised within four months, enabling the masters to overcome a spinners' strike. In later years, having lost commercial confidence as a financially sound manufacturer, he was deprived of his business and died poor. Yet in his 60th year, after a strike by the workmen on the Conway tubular bridge, Robert Stephenson had him produce a "Jacquard" machine on the pricked-card principle of the French loom for punching holes of any pitch or pattern in bridge and boiler plates.

Machine for Making Dead Eyes.

Invented by Mr Brunel.

In order to improve their fortunes, the landowners of the hilly South Durham coalfield needed a link with the sea. A canal was proposed, but surveys showed that the route was too hilly; instead a horse-worked wagonway was built between the coalfields and Darlington. Later a railway from Darlington to Stockton was surveyed and engineered by George Stephenson, who envisaged the use of steam locomotives upon the much improved rails. These were of the 4 foot $8\frac{1}{2}$ inch gauge (taken from Stephenson's earlier wagonways), which was to become the worldwide "standard" gauge. Only one steam locomotive was ready at the line's opening in 1825 – Stephenson's *Locomotion No. 1* – but others followed and despite some early problems, when they proved to be under powered, the principle of steam traction was at last established. It showed a saving of 30 per cent on the haulage cost per ton mile compared to horse haulage. The Stockton and Darlington Railway was the world's first fully public railway line and the first to utilize steam to haul passengers on a regular basis.

Exorbitant charges and long delays for their canal shipments between Liverpool and Manchester led some of the local businessmen to propose a railway to break the canal's monopoly. George Stephenson surveyed the route, and suggested either steam locomotion or cable haulage; cableways, which he had used at various coal mines were practical because of the severe inclines at the approach to the Liverpool terminus. It had not yet been established that steam could operate on grades as steep as one in 100, but it was agreed that trials should be arranged to see if a suitable locomotive design could be found to operate the line throughout. These took place at Rainhill in 1829, with five "runners," one of which was a horse-worked contraption. Of the four steam locomotives, *The Rocket*, designed by Stephenson's son Robert, proved its reliability and haulage capacity decisively, ending the trial with a demonstration on a one in 100 gradient. *Rocket*'s multi tube boiler design, and consequently its ability to produce sufficient steam, set the pattern for future locomotive development, and the Rainhill trials marked the true commencement of the steam locomotive's long reign of supremacy as land transport motive power.

Richard Trevithick had operated his *London Road Carriage* at speeds up to 10 mph in the streets of London, but had lost interest because he could not find a financial backer. Not until the 1820s did the idea of steam road carriages attract backers, and by then a number of inventors were at work on new designs. Some believed that the poor state of the roads would not allow the steam-driven wheels sufficient adhesion, and one such man, Sir Goldsworthy Gurney, patented in 1825 a rotary "walking" device which dug spurs into the surface of the road as it turned. Two years later he pinned his faith to the wheel and produced a 21-seater coach. This had a high-pressure superheated steam boiler of the water-tube type, working two 9-inch bore by 18-inch stroke cylinders driving the rear wheels by connecting rods and cranks on the rear axle. After trials in London, a special demonstration run from London to Bath and back was arranged. The outward journey ended in failure. After repairs, the coach steamed off to London, only to be attacked by an angry mob of agricultural workers at Melksham fair. Gurney was seriously injured, but not disheartened by the episode, and later developed the idea of light, high-speed steam tractors able to tow a wagon or coach.

Walter Hancock developed a vertical boiler for steam road carriages, which worked at high pressures and had a number of chambers above the fire grate, connected by tubes at top and bottom. A fan on top of the firebox provided a forced draft. His first carriage, a small three-wheeler, used two oscillating cylinders, driving directly onto the front axle; later he refined his method of drive and improved the boiler further. His *Infant* of 1831 had a 10-seater charabanc body, and Hancock inaugurated the concept of the fare-paying regular road passenger service, over common roads, between Stratford (East London) and the City. Like so many inventors of the period, Hancock suffered at the hands of unscrupulous financiers who attempted to plagiarize his boiler design, but failed. His later and larger carriages, such as the *Automaton* of 1836, were very successful, and featured Samuel Miller's artillery wheels, to give a better ride. Hancock personally operated *Automaton* between Paddington and the Bank and Moorgate, making over 500 trips and carrying some 12,000 passengers without mishap. General development of the steam carriage ebbed after 1840, when the steam railway came into its own, and the horse once again ruled the streets until the coming of the electric streetcar and the automobile.

The camera obscura has been known for a thousand years, and was used by the Arab scientist Alhazen of Basra to observe eclipses of the sun in the 10th century. In a dark room with a small hole cut in the center of one wall, straight lines of light from a brightly illuminated scene pass through the hole to register an image upside down on the far wall.

By the early 19th century the camera obscura had been made small and portable. A mirror was set inside at an angle of 45 degrees, and the image was projected onto a screen of frosted glass, to be hand-registered if required by putting tracing paper on the glass. Between 1796 and 1802 Thomas Wedgwood attempted to secure permanent records from the camera obscura by placing on the glass a sheet of paper moistened with silver nitrate. His object was to portray natural scenes, stately homes, and details of flowers and plants to be transferred to Wedgwood porcelain as decoration for dinner plates. Wedgwood could not fix the image or stop the action of light on the sensitized paper, and he abandoned the project.

Twenty years later, at Châlon-sur-Saône in France, Joseph Nicéphore Niepce achieved a permanent image from the camera obscura, on which he had been working for some years. He received the image on a plate coated with a varnish of bitumen and oil. After an exposure of eight hours, he bathed the plate in acid, which dissolved the varnish under the dark areas of the image (where, less influenced by light, it was still soft and soluble). He then sent the plate to the engraver LeMaître who deepened the lines, inked the plate, and printed the resulting picture like an engraving. The first existing photograph utilizing this method dates from 1826, when Niepce began to use pewter for his plates. The image shows a view from Niepce's attic window. Because of the eight-hour exposure, the sun lights both sides of the street.

At LeMaître's suggestion Niepce adopted a copper sheet thinly plated with silver as the best medium for etching a picture from which he hoped to print many copies. He later developed the use of a glass plate, but as a transparency, never as a negative for further prints. In 1829 Niepce, 64 years old and impoverished, went into partnership with Louis-Jacques-Mandé Daguerre, a hustling entrepreneur aged 42. Niepce took little part in the exploitation of his methods and died shortly after. Daguerre himself had a rocky decade before he made his fortune in 1839.

During that time William Henry Fox Talbot had been striving to capture permanently the images he received with his camera obscura. Following Wedgwood, he used paper impregnated with a silver salt. But, before soaking the paper in silver nitrate, he had first put it in a bath of common salt and then dried it. As a result, when dipped into silver nitrate solution, a chemical change took place and the paper was impregnated with light-sensitive silver chloride. Fox Talbot exposed these paper sheets in cameras placed in his home, Lacock Abbey in Wiltshire, and after half an hour washed the sheets with salt solution, and later with potassium iodide. He learned from Sir John Herschel, the astronomer son of Sir William, to fix the images on the paper with sodium hyposulphite, which dissolved away the remaining silver salt. One of his earliest pictures, exhibited by Michael Faraday at the Royal Institution in London, was an inch-square paper print of a lattice window at Lacock Abbey made in August, 1835.

Fox Talbot progressed to making transparent "negatives" from his camera by rendering the paper translucent with wax, and contact-printing these negatives by sunlight acting on paper impregnated with silver chloride.

Talbot announced his development of photography prematurely in January, 1839, at a time when Daguerre, after 10 years' gestation, was about to deliver in Paris. Daguerre had started working life as a theatrical scene painter; he had a notable flair for pictorial composition, and some of his easel paintings are still greatly admired. He had developed the Diorama, a peepshow combining three-dimensional figures with painted backdrops, as a commercial venture. Meanwhile, he strove to make a working success of his interest in Niepce's photography. After many years he used mercury vapors to develop images on a sheet of silver-plated copper, previously sensitized by exposure to iodine fumes, and protected within a light-tight holder until exposure. The plate was then fixed with salt (later hypo) and washed clean of chemicals. The result was a positive Daguerrotype on polished copper. In 1839, the year when Daguerre made a carefully staged announcement of his process and reaped an immediate fortune, he produced a striking picture of a Paris boulevard with a pedestrian having his boots cleaned. It was the first photograph of a human being, but in its composition as well-chosen as many a painted Impressionist street scene.

Science responds to patronage, though it does not depend on it. King George III had enthusiastically appointed William Herschel as court astronomer and excitedly demonstrated Herschel's giant telescope to the Archbishop of Canterbury with the assurance *I will show you the way to Heaven*. George IV, always a flamboyant patron of art and literature, recognized science only twice – by contributing to a statue of James Watt and by instituting the annual award of two gold medals for papers to the Royal Society on scientific subjects. By coincidence, during his reign, a number of advances were made in computation, communication, and optics, which lay in obscurity for many years. In 1822 Charles Babbage began work on the construction of an *analytical engine*, a calculating machine with a network of levers, wires, and gears that anticipated the modern computer. This was no lunatic project. Babbage, then 40 years old, was already a fellow of the Royal Society with a reputation for having fathered a belated revival of British mathematics, initially at Cambridge, where the specialization took root. He had a clear, if forceful, intellect and he was a man of some means who spent £6,000 on constructing a model that would add six-figure numbers, then persuaded the Government to put up money for a far more elaborate machine with a memory based on punched cards and print-out facilities. Personality difficulties and construction problems made progress slow and, though after 30 years one machine was built and used to cast actuarial tables, the project was not successful: Babbage wrote his perfected logarithm tables without it. Throughout his long life, Babbage brimmed with ideas. He was an early advocate of minute specialization of labor, and his cost analysis of letter-carrying convinced the British Government to pioneer the flat-rate *penny post*. Among many innovations, he made the first opthalmoscope to examine the retina.

In 1820, Hans Oersted of Copenhagen published his discovery that an electric current would deflect a magnetic needle. Michael Faraday in London experimented to convert the electromagnetic motion which Oersted had observed into continuous mechanical movement. In 1821 he rigged up an arrangement by which an electric current made a conducting wire rotate around a fixed magnet while a pivoted magnet rotated over a fixed conducting wire. He had demonstrated in principle the first electric motor – the beginning of his work in electromagnetic induction made the motion of a conductor in a magnetic field produce an electric current. He thus constructed the first generator, a dynamo converting mechanical energy into electrical energy. It was a progressive experiment which began with Faraday winding a wire coil around one sector, about half the circumference, of an iron ring. He switched current through the coil and confirmed that a magnetic field was created in the iron ring. He then wound another coil around the other half of the ring to see if the magnetic field of the first coil would induce an electric current in the second. It did, but the induced current was only momentary, occurring just when the current of the first coil was switched on, and switched off. He deduced that the current occurred in the second coil only when the magnetic force sprang out of the first (when it was switched on) and when the magnetic field collapsed in the first (when it was switched off). In order to secure a lengthy series of "on" and "off" excitations in a magnetic field he mounted a copper disc between the poles of a powerful bar magnet and rotated the disc. A steady electric current was generated between the rim of the disc and its axle. This laboratory generator of alternating current was only slowly adapted for commercial use.

In 1836 William Fothergill Cooke, a 30-year-old retired Indian Army officer, began experiments with electric telegraphs. Realizing his fragmentary technical knowledge, he consulted Michael Faraday, who recommended him to Charles Wheatstone, professor of natural philosophy at King's College, London. Wheatstone, four years Cooke's senior, was already experimenting and joined Cooke in partnership. In 1837 they devised a telegraph which Robert Stephenson installed for signalling along a short section of railway in London. Cooke and Wheatstone had begun with a diamond-shaped board with five magnetic needles pivoted across its breadth. The needles could move to left and right, and two needles were moved concurrently by the operation of a sort of piano key for each letter. The point where the needles intersected on the reception board indicated the letter that had been transmitted. In 1843 a board with only two needles was substituted, requiring the adoption of a code rather than alphabetical letters. Within two years of this improvement there were over 1,000 miles of telegraph operating on English railways, and the railway station became the conventional public telegraph office.

Samuel Finley Breese Morse was born in Boston, Massachusetts, in 1791, the year of Chappe's first semaphore telegraph. In his 40s he became enthusiastic about electric telegraphy and began working with Joseph Henry, who was already an expert in electromagnetism, and who freely aided Wheatstone. In Paris in 1839, Morse swapped trade secrets with Daguerre, the pioneer of photography, and later in the United States he produced early daguerrotypes – without permitting Daguerre to meddle in France with electric telegraphy. Morse annexed Henry's theoretical work and marketed it ruthlessly, while Henry cheerfully became the first secretary of the Smithsonian Institution. The first Morse telegraph line for public business was opened in America in 1845, though Washington had been linked with Baltimore for official and experimental purposes in the previous year. Sole credit is due to Morse for the invention of his code – dots and dashes made by energizing an electromagnet – which was devised to be *read* from a paper ribbon, and only later began to be interpreted by operators aurally.

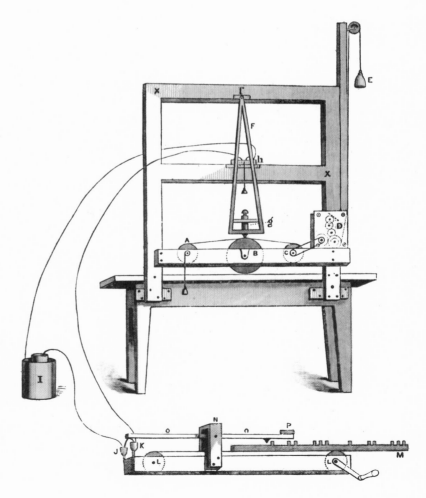

Eli Whitney was not notably richer from his gin which, though it speedily increased cotton production in the South tenfold, had not yielded the collectable royalties it had truly earned. Whitney therefore staked his future in arms manufacture, and in 1798 he had secured a government contract to provide 10,000 muskets. He could produce such a vast quantity only by developing the principle of interchangeable parts, which in turn demanded a new precision in machine tools. Only by accurate machining could he overcome the slowness and variability of hand-working where every part in a composite device was a fine fit for the assembly it was shaped for, but was unreliable as an alternative part for another, supposedly identical product. Whitney therefore concentrated on precision engineering and the rationalized specialization of labor. As early as 1801 he had been able to throw down the parts of disassembled muskets and build a weapon from a random choice of constituents. By 1820 he had produced the earliest existing American precision milling machine, with an automatic feed drive worked from a rotating lead screw.

John George Bodmer was concerned with motion of a different nature: the secure, fast, efficient consignment of a complicated workpiece from one process to the next. This is the principle of modern factory automation. Bodmer was a prophet without much honor in his own country, Switzerland; and in Great Britain he was esteemed but not enthusiastically followed. In the less hidebound United States he was appreciated by the more open-minded manufacturers who were chronically short of laborers. After leaving the millwright's shop which he had set up in Zurich, Bodmer at the age of 20 had expanded a Black Forest textile machinery factory to take in the production of small arms. Almost contemporaneously with Eli Whitney he introduced the preparation and assembly of identical interchangeable parts. As an armaments officer for the Grand Duke of Baden, Bodmer tried to apply the mass-production principle to field artillery. He installed the belt conveyor, but timing was against him, for the battle of Waterloo ended massive trade profits from war in Europe for a long time. A year later, in 1816, he visited England and toured the principal machinery shops, textile mills, and ironworks. He recognized English mechanical pre-eminence, and in 1824 set up a small textile machine factory in Bolton — now notable for installing the first known traveling crane for the transit of workpieces. Between 1833 and 1841 Bodmer established a textile machine factory in Manchester, designing almost all its machines himself, and in two embracing patents of 1839 and 1841 registered 40 inventions. His layout and system were outstanding by modern standards. He had traveling cranes and railroad trucks serving the length of his shop, and he innovated the chain grate for boilers, which anticipated automatic stoking. His standards of tool design were revolutionary. His vertical boring mill, which he called a circular planer, was, along with his automation system, ignored in England but adopted in America. His most speedily accepted innovation was the milling machine, one of his 1839 patents.

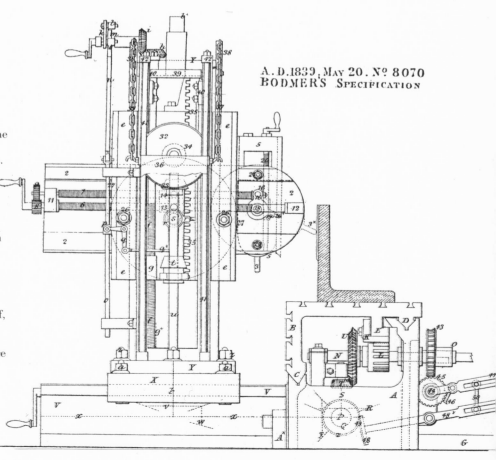

A.D. 1839, MAY 20, Nº 8070
BODMER'S SPECIFICATION

Milling involves cutting metal from a workpiece by a rotating edge. Vaucanson had used a primitive machine for cutting gear wheels of clocks. Development of the milling machine was again keenly pursued in the United States but long ignored in England, although, before Bodmer, James Nasmyth had made a specialized miller for machining the flats of hexagonal collar nuts. It was the first machine to have a continuous coolant – a dripping can of water. When he built his milling machine, Nasmyth was a 22-year-old gentleman apprentice to Henry Maudslay. In the previous year, 1829, he had devised a coiled spiral spring for drilling in otherwise inaccessible parts of a workpiece. This was frequently re-invented, and in Nasmyth's old age, when it housed a dentist's drill, it was proudly presented to him by his own dentist as the latest American innovation. On the death of Maudslay, Nasmyth set up his own shop, first in Edinburgh but more profitably in Manchester at the Bridgewater Foundry. In 1836 he produced an important shaping machine in which the cutter moved back and forth across a stationary workpiece; it grooved with the operation of a cold chisel, and had potential to shape any small surface founded on straight lines.

Soon afterwards he was consulted professionally regarding the forging of a paddle shaft for Brunel's proposed new steamship, the *Great Britain*. The shaft, intended to be 30 inches in diameter, was too large to be forged by the only tool then available, a trip hammer falling a comparatively short distance with the arc flight of a man wielding a hammer. Nasmyth was a dedicated doodler, and he set his mind to the problem. *In little more than half an hour I had the whole contrivance in all its executant details before me in a page of my scheme book*, he later claimed. It was his famous steam hammer of 1839, an inverted steam-engine cylinder with the hammer head (soon to weigh four tons) fitted to the end of the piston rod and falling vertically. At first it was only raised by steam and dropped by its own weight. Soon he applied steam to the descent, greatly increasing its power, and creating his specialized pile driver. Yet the modified steam hammer was capable of delicate control. It could be made to fall, according to the catalog of the Great Exhibition of 1851, *with power only sufficient to break an egg shell*. Because the paddle shaft for the *Great Britain* was canceled in favor of a screw-propeller shaft, Nasmyth did not bother to patent the steam hammer. But Schneider, the French ironmaster of Le Creusot, saw Nasmyth's sketches and built his own hammer in 1841. One of Nasmyth's last professional duties as an engineer was to preside over the Royal Commission on Small Arms of 1853 which recommended the adoption of the American system exploited by Colt for the production of the Enfield rifle. He retired at the age of 48 and devoted himself to astronomy.

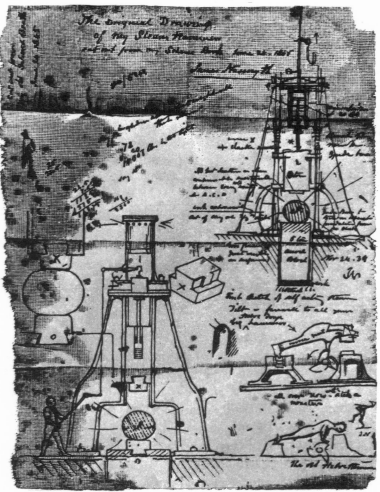

The screw propeller as a means of driving a steamship was really the discovery of two independent inventors. In 1837 Sir Francis Pettit Smith was experimenting with a six-ton canal boat which he had equipped with an "Archimedean" screw propeller. The propeller was made of timber and its corkscrew-like shape was two full turns in length. During the trials, pieces of the screw broke off, leaving, in effect, two separate blades and producing a considerable improvement in performance. Smith was so encouraged by this that he went on to design the successful *Archimedes*, using a twin-bladed screw. At the same time, a Swedish ex-Army officer named John Ericsson was experimenting with a screw propeller consisting of a double set of fan-shaped blades held together with rings; in the original design the two sets were contrarotating. The first application of this screw was in the vessel *Francis B. Ogden*, which ran trials on the River Thames, and Ericsson went on to design the highly successful screw-driven *Monitor*. Although Smith's *Archimedes* convinced Brunel to use the screw propeller, it was a six-bladed screw akin to Ericsson's design which he was to install in the *Great Britain*.

The little cross-channel paddle steamer *Sirius* (703 tons) was the first vessel to make a trans-Atlantic crossing entirely under steam power. At the time, the ports of London, Bristol, and Liverpool were all eager to capture any trade which might develop from the use of steamships on the North Atlantic crossing, but it had yet to be proved that such a crossing was practical. Construction of Brunel's *Great Western* for the Bristol–New York route was already well advanced when Curling and Young of London laid down the *British Queen* as their contender. As it became clear that *Great Western* would be ready long before their ship, the London group chartered the ship *Sirius* from the Anglo-Irish service. *Sirius* had been launched in 1837 and although new, required modification and increased bunker space for the Atlantic crossing.

Sirius sailed from London on March 28, 1838, with 40 passengers and a crew of 35. Although the *Great Western* set out in pursuit a few days later, *Sirius* was able to refuel at Cork and still retain the lead. With 3,000 miles gone, coal began to run short, but despite popular legend that the ship's furnishings had to be burnt as fuel, it in fact reached New York after 19 days with 15 tons of coal still remaining. The near miss did little to prove the practicality of steam power on such a crossing, but it received a rapturous welcome from the people of New York. The *Great Western* arrived soon after, having crossed in 15 days with 200 tons of coal still unused.

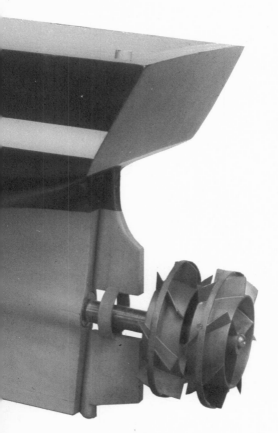

192

The three-masted topsail schooner *Archimedes* was launched at Millwall in 1838 and carried a steam engine built by the Rennie brothers. It was the first successful vessel to be driven by the screw propeller designed by Sir Francis Pettit Smith, which had first been used unsuccessfully in trials of a launch on the Paddington canal. Brunel's associate Thomas Guppy "sailed" to Liverpool in the *Archimedes* after it called at Bristol in 1840, and was so impressed that he convinced Brunel to charter the *Archimedes* for six months of trials. Brunel's exhaustive experiments revealed a number of faults in the design of the ship but he determined to install screw propulsion in his new ship, the *Great Britain*. The engine of the *Archimedes* drove the propeller shaft through a straight gear drive, and although it gave the 237-ton ship a speed of nine knots, it was noisy, inefficient, and produced exceptional vibration. Nevertheless, as an experimental ship *Archimedes* proved the superiority of the screw over paddle wheels and set the stage for the first commercial application of the screw propeller.

THE "GREAT BRITAIN" STEAM-SHIP, NEWLY RIGGED.

The *Great Britain* had been a continuation of the ambitious scheme of the engineer Isambard Kingdom Brunel, who designed the Great Western Railway in England, to *make the railway longer* by extending its range from Bristol to New York. He built the wooden-hulled paddle steamer *Great Western* which made a 15-day maiden voyage to New York in April, 1838. He proposed to follow this with a bigger ship made of wrought iron, originally to be paddle-driven but finally built with a propeller shaft. The *Great Britain* was launched by the Prince Consort from Patterson's yard at Bristol in 1843, and the scene was recorded by Fox-Talbot in one of his early historic photographs. After an inauspicious first winter aground off Ireland, the *Great Britain* – 320 feet long with a weight of 3,443 tons and displacement of 2,984 tons – began 30 years' service by carrying 60 passengers and 600 tons of cargo from Liverpool to New York in 15 days in 1845. The eventual success of the ship accentuated the use of wrought iron, which had the same load-carrying capacity as twice the weight of cast iron.

On March 24, 1843, J.A.Roebuck, Member of Parliament for Bath, introduced into the British House of Commons a motion which would lead to an Act of Incorporation of Henson's Aerial Transport Company. This company was to operate an Aerial Steam Carriage to convey passengers, goods, and mail through the air. Four days later a patent was granted; within the week the *Mechanics Magazine* published the full specifications. A sheaf of finely executed illustrations was distributed, showing the Aerial Steam Carriage flying over London, Paris, the Pyramids, India, and China. Simultaneously, the inventor, William Samuel Henson, was soliciting public investment in the promotion of a company that would run Aerial Steam Carriages in fleets *to convey Passengers, Troops and Government Despatches to China and India in a few days*. The idea of transporting troops by air, put forward almost a century before its actual inception, was a masterstroke of promotion in those imperial days. The aircraft it proposed to use for this great venture had not even been built, let alone tested, but it was described in great detail. Henson was, in fact, an enthusiastic, but not very profound inventor, then aged 38. In Chard, Somerset, where he ran a lace factory, he had registered several inventions during the previous eight years for lace machinery and an improved steam engine. His aircraft design, begun after he had turned to aeronautics in 1838, was influenced by Cayley; but Henson introduced original features that were influential for 60 years. Ariel, the Aerial Steamer, or the Aerial Steam Carriage, as it was variously called, was a monoplane with rectilinear wings of 150 feet span and 30 feet chord that were covered with oiled silk. They made a single plane, without dihedral setting, and were cambered and double-surfaced, which is not obvious in some of the illustrations. They were ribbed, and braced with wires tautened by king posts. A fan-shaped tail plane could be diminished in spread – its maximum area was one-third of the main plane – and it worked as an elevator by being moved up and down from a pivot. A vertical sail beneath the tail acted as a rudder. This was, in historical design terms, the first separation of the elevator and rudder, which had been cruciform in conception from Leonardo to Cayley.

A 25 to 30 horsepower Henson steam engine of special efficiency and lightness, housed in the fuselage, worked two six-bladed pusher airscrews mounted behind the main plane. There was a tricycle undercarriage with tension wheels, adopted from Cayley's innovations. Because it was calculated that the power of the engine would be sufficient to maintain the Aerial Steamer only in level flight, Henson intended for the airplane to take off downhill. The gross weight was 3,000 pounds, including a 600-pound engine, which meant that the 1,000-pound payload would hardly have accommodated six soldiers with their equipment. The airplane was never built because the public failed to subscribe to Henson's company, though they accepted that the Air Age was near at hand and that the future Aerial Locomotive would resemble Henson's creation. The Aerial Steam Carriage incorporated all the aeronautical features (except dihedral wings: and ailerons, which had not yet been dreamed up) which most of the early practical monoplanes of the 20th century were to adopt. Lacking the capital to build his fleets, Henson managed, with the co-operation of John Stringfellow, to construct a model with a wingspan of 20 feet and a miniature steam engine. In 1844 it achieved one solitary "power glide."

After Sir William Congreve had brought 19th-century military rocketry to its zenith, a British engineer, William Hale, created a new design which eliminated the stick in the rocket, stabilizing and guiding the missile by three metal vanes set to catch jets from the exhaust vent and rotate the rocket. Hale also designed riveted steel bodies, permitting a pressure of up to 23,000 pounds per square inch and a range of a mile and a quarter. Among Hale's clients was the U.S. Government, which used 16-pound Hale rockets effectively in the Mexican War of 1846–1848, and ineffectively in the Civil War. Because rocket launchers were lighter and more mobile than cannon and mortars, Hale rockets continued to be used in mountains and swampy terrain, particularly by the Dutch in "pacifying" the Celebes. Their last effective use was made by the Russians in the war against Turkestan in 1881. Their successors in the next century were to be the guided missile batteries.

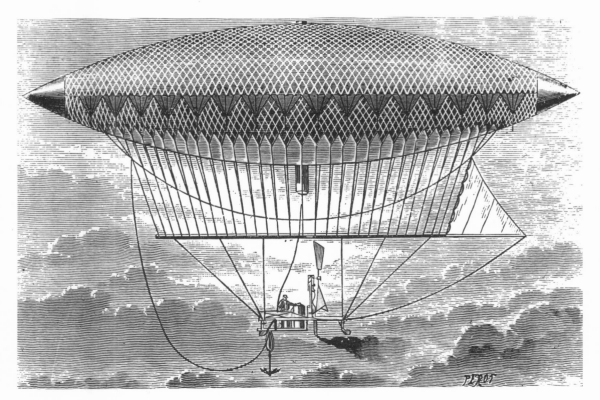

By the age of 27, Henri Giffard had already made a fortune from his invention of the steam injector and had designed a steam-driven helicopter. On September 24, 1852, he flew the first successful "navigable balloon," or airship, 17 miles from Paris to Trappes. The envelope, inflated with coal gas, was 44 meters long with a maximum diameter of 12 meters and a capacity of 2,500 cubic meters. Giffard produced a three-horsepower steam engine, slung 20 feet below a 75-foot boom keel, well removed from the flammable envelope. It drove a three-bladed airscrew with an 11-foot diameter at 110 revolutions per minute to propel the aircraft at five miles an hour, but it did not produce enough power for a circular turn in wind, and the sail used as a rudder was ineffective. In 1855 Giffard crashed in a similar airship, but with no serious injury.

Though dubbed in his time the "Father of
Aerial Navigation," Sir George Cayley only
recently won international recognition as the
inventor of the modern airplane and the
founder of the science of aerodynamics.
The machine roughly sketched here by Cayley
is a true airplane, i.e. a heavier-than-air
construction supported by the dynamic
reaction of air flowing about a fixed wing.
This was Cayley's first contribution to
aeronautics – isolating the flapper for
propulsion and introducing the fixed wing,
which settled the problem of lift and made
thrust a separate consideration, instead of
combining the two functions as previous
birdmen had done. Cayley never achieved
adequate thrust because he could not devise an
engine light enough in weight and paid
insufficient attention to the conception of the
airscrew. In this machine for propulsion he
installed flappers to be worked by the pilot
manually through levers in the gondola.
Technically, therefore, this airplane is a glider
with auxiliary propulsion: it flew, manned but
not piloted, in the spring of 1849. Cayley
recorded: *A boy of about ten years of age was
floated off the ground for several yards on
descending a hill, and also for about the same
space by some persons pulling the apparatus
against a very slight breeze by a rope.* In 1853,
the date of this sketch, Cayley launched his
"New Flyer" which, manned by his coachman,
glided independent of ground control for
several hundred yards. Cayley's documented
research and achievement was partially known
to, but in general ignored by, his 19th-century
successors; detailed specifications of this Old
Flyer were not to be published until 1961.

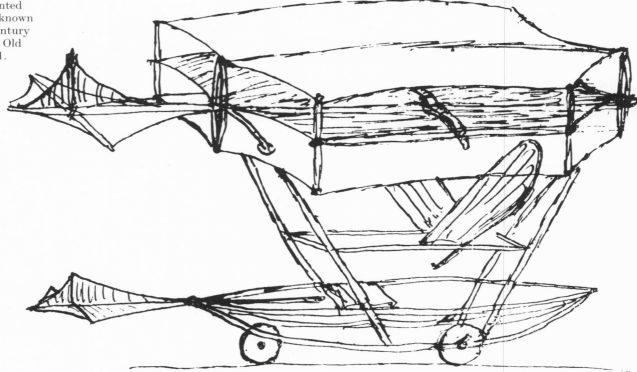

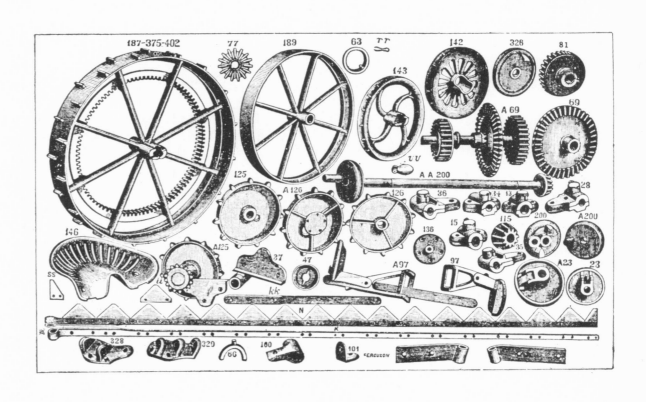

In the 1850s two groups of English technicians who visited the United States reported that Americans were producing a wide range of products by more highly mechanized and more standardized methods than could be found in Europe. About the same time, the arms factory at Enfield (England) re-equipped itself with machine tools produced in Vermont. By the early 1880s an Italian observer commented upon the *spirit of inventiveness which is so extraordinarily developed among American workers to whom our age is indebted for its more useful innovations and discoveries.*

The growth of the United States in the 19th century opened a new chapter in world history. In more than one way the United States was never an underdeveloped country. From the very beginning of their history the American people were highly literate, possessed of the advanced skills they had brought from Europe, and could count on an extraordinary abundance of natural resources. The scarcity of labor relative to the abundance of other resources, the rapid growth of an economy with an expanding frontier and a continuous inflow of population, mostly in the active age groups, the absence of those old traditions and vested interests which imposed severe restraints on new technologies in Europe – all these were factors highly favorable to a high level of inventiveness.

Overleaf: American mechanical reaper spare parts, 1850

Social and cultural factors must also be considered. The early immigrants came from England – the country which had most enthusiastically embraced Francis Bacon's utilitarian view of knowledge. The Puritans in particular felt the urge to be practical and active. Almost obsessively, they insisted upon the moral value of expending physical energy in the handling of material objects. In their single-minded outlook there was no room for contemplation, which they identified with idleness. Moreover, the origin of their new society was inimical to the classical culture which had considered leisure the only possible source of theoretical knowledge. For the sons of the peasants who had fled Europe to escape hardship and exploitation, the idea that knowledge should be the fruit of inherited leisure was simply inconceivable. Instead, they viewed theoretical knowledge as the fruit of common toil, which must be directed not to self-gratification but to providing for human needs.

The strengths and the limitations of the American outlook were described in an article in *Putnam's Magazine* in 1854: *The genius of this new country is necessarily mechanical. Our greatest thinkers are not in the library, nor the capitol, but in the machine shop. The American people is intent on studying, not the hieroglyphic monuments of ancient genius, but how best to subdue and till the soil of its boundless territories, how to build roads and ships, how to apply the powers of nature to the work of manufacturing its rich materials into forms of utility and enjoyment. The youth of this country are learning the sciences, not as theories, but with reference to their application to the arts. Our education is no genial culture of letters, but simply learning the use of tools.*

The accent now was on managing things, not on ordering lives. But was *learning the use of tools* all that there was in life ? The pioneers were not troubled by such doubts.

By the first quarter of the 19th century a farmer's average agricultural yield had been increased considerably by better ploughing, better sowing, better hoeing, and a slow progress in farming theory and technique. The gain in output only exaggerated the inefficiency of the harvest time. Special labor had to be hired for the harvest, which demanded more work per unit of production than any other agricultural task. There was a certain flexibility in allocating pickers for root and fruit crops. The critical crop was always the all-important grain harvest, when weather conditions cross-acting on ripeness often required emergency action. Eighteen centuries earlier, the Gauls had experimented in that field. The elder Pliny described a box on wheels which an ox pushed into standing grain while static knives cut the stalks at head height, so that the ears fell into the cart. In 1826 Patrick Bell, a 27-year-old Presbyterian minister from Carmelyie in Scotland built and demonstrated a similar machine, pushed by two horses into standing barley. The grain was then swept by forward sails against rather crude scissors which snipped the stalks and deposited them to the side from an inclined platform. At this time, in the United States, another Presbyterian, Robert McCormick, was striving to construct his own mechanical reaper. He had started in 1816, using the same pushing principle, but his fixed sickles failed to cut the grain as it was flung at them by the revolving laths. He modified and adapted, always unsuccessfully, for 15 years, by which time his eldest son Cyrus Hall McCormick was 22 years old. In July 1831, Cyrus demonstrated his own mechanical reaper. It was pulled into the standing grain by a single horse, harnessed to the right hand of the machine, which was to be the stubble side. There were no mechanically moved scissors as in Bell's contraption (which Bell soon abandoned for McCormick's principle) but a straight knife or cutter bar with serrated edges moved reciprocally by gears from the main wheel. The grain was held against the knife by a revolving reel, also operated by the main wheel, which laid the cut stalks onto a platform, which a man cleared periodically. Finger guards extended from this platform forward

to prevent the grain from slipping sideways while being cut – the problem which Cyrus' father had never overcome. A divider on the left side of the machine separated the grain to be cut from that left standing. McCormick used his reaping machine privately, and gradually improved it. He patented it in 1834 and produced his first models for sale in 1840. Sales were slow until, in 1847, he began manufacture in Chicago. By 1849 he was employing 120 men and selling 1,500 machines a year. McCormick was 42 years old when the Great Exhibition was held in Hyde Park, London, in 1851. There, he took all the honors in field trials against rivals. Queen Victoria noted his machine in her diary, and the London Press commented ecstatically: *Mr McCormick's reaping machine has convinced English farmers that it can be economically employed . . . In agriculture, it appears that the machine will be as important as the spinning jenny and power loom in manufacture.* Two further innovations, both American, clinched the success of the McCormick reaper. Even before Isaac Singer pushed sewing machines into the homes of Europe and America, Cyrus McCormick was selling his $125 reapers to farmers on easy terms – $35 down plus carriage, and the balance over 18 months. The other great advantage was the availability of spare parts. Because of mass production, any one of 35 interchangeable parts could be bought for replacement. McCormick went on making money and using it to endow professorships at a theological seminary, buy the *Chicago Times*, and develop what became the International Harvester. McCormick began an agricultural revolution which, in a century and a half, cut the man-hours necessary to produce a bushel of corn by almost one hundredfold. This was manifestly not all his own work. But in his own time, he was revolutionary. Professor Reynold M. Wik declares: *The reaper was probably the most significant single invention made in agriculture prior to the Civil War, for it reduced the harvesting labor by one half at the critical point when work had to be completed quickly to save the crop.*

At the Great Exhibition in Hyde Park, London, in 1851 European engineers visiting the Crystal Palace were astonished by implements which had been manufactured in the United States. Produced on a variety of special-purpose machines, each of these implements had been designed for limited functions to turn out identical parts under comparatively unskilled supervision for subsequent assembly. The products were made with enviable precision and high practical efficiency. In field trials on agricultural machinery the McCormick reaper beat all comers. The system of standardized production and assembly of interchangeable parts – not yet incorporating the moving belt later borrowed from the Chicago slaughterhouses – was promptly dubbed "the American system of manufacture." Apart from agricultural machinery, the exhibits most recognizable for their precision and reliability were small arms: the rifles of Robbins & Lawrence and the revolvers of Samuel Colt.

Colt had only one good inventive idea in his life, but he exploited it with a peculiar genius for selecting the best productive methods he observed in other plants, and a flair for good relations with his workmen. He was born in Hartford, Connecticut, in 1814, and ran away to sea when he was 16. Whittling wood in mid-ocean, he carved a revolver, a hand gun with a cylinder which held six cartridges and was rotated by cocking the hammer. He patented this in 1835, and within two years his revolvers were being used with success by U.S. troops against the Seminole Indians in Florida. The Indians were pacified, business slackened, and Colt's factory went bankrupt. He was recalled to action by a Government order for 1,000 Colt revolvers to be used in the 1846–1848 war in Mexico, which added California and Nevada and other territory to the Union. Colt placed the order with Whitney's armory at New Haven, where the American system was already in operation.

He was so impressed with the efficiency of the new system that he introduced it, in a greatly developed form, when he built his new armory at Hartford in 1853 under the management of Elisha K. Root, a highly skilled mechanic who designed dozens of specialized machine tools with the objective of entirely eliminating the manual fashioning of parts. The same year Colt established a Thamesbank factory in Pimlico, London, and, employing unskilled labor with American supervisors, was soon making 1,000 revolvers a week. In 1854 he exported the American system in its entirety to England by winning a contract for its installation in the new Government small-arms arsenal at Enfield. Both enterprises were speeded and made highly profitable by the onset of the Crimean War of 1854–1856, and there is no doubt that Colt demonstrated the urgency of rapid arms manufacture by first supplying his revolvers to the Russians, who were Great Britain's enemies in that conflict. At the end of the war, having sold 35,000 revolvers, Colt dismantled his London factory.

Even in England it had been the armaments industry which pioneered machine-tool design. In 1809 the block mill at Portsmouth Dockyard was producing complicated tackle for British warships, using machines which cut the labor previously required by 90 per cent. American engineers made their first significant contributions to the development of the lathe when they began to design special-purpose machines for interchangeable parts for arms assembly. In 1818 Thomas Blanchard built his patent lathe for turning gun stocks: the first lathe turning irregular forms, operating on the pantograph principle using a master gun stock.

A metalworking machine tool was the logical development. The pioneer was devised to meet the requirement of a vast number of identical screws for the percussion locks of 30,000 pistols ordered by the U.S. Government. For this purpose, Stephen Fitch of Middlefield, Connecticut, designed and built the first turret lathe in 1845. The long cylindrical turret revolved on a horizontal axis and carried eight tools mounted on spindles, each of which could be advanced as required, with the operative tool uppermost. A three-armed capstan advanced the turret carriage and applied the feed. In this way, eight successive operations could be rapidly performed without stopping the machine to change tools. Stephen Fitch was recognized by his fellow engineers as *the creator of one of the most time-saving machines ever invented.*

The armaments industry next called into being the universal milling machine. Milling is the removal of metal from a workpiece by passing it beneath or across a rotary cutting tool. The French clockmakers used a specialized milling machine in the 18th century for cutting gear wheels, but it was the Americans who produced genuinely versatile milling machines. The culmination of a series of designs by Frederick W. Howe, working for the Robbins & Lawrence rifle manufacturers, was taken over by Elisha Root, the masterly mechanic whom Colt had picked to run his factory. Root, with his assistant Francis Pratt, improved on a machine which Howe had built for delivery to the British Government's Board of Ordnance for the new factory being built at Enfield. The Root/Pratt model was built at Hartford in the Phoenix Ironworks of the George S. Lincoln Company, and it became known as the Lincoln Miller. This robust production machine was sold by the thousands throughout the world. Its impact was tremendous: 100 universal millers put into service at the Colt Armory doubled the production of the plant. L.T.C.Rolt has said of it: *no other tool contributed more to the early history of the American system and yet the universal milling machine was created by that system.*

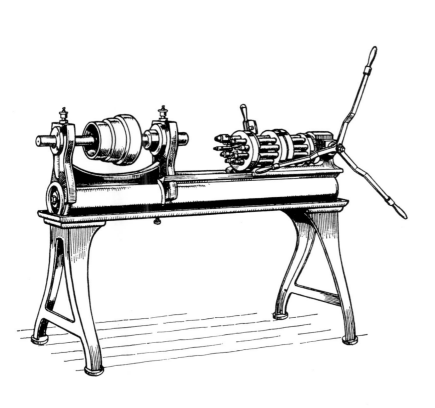

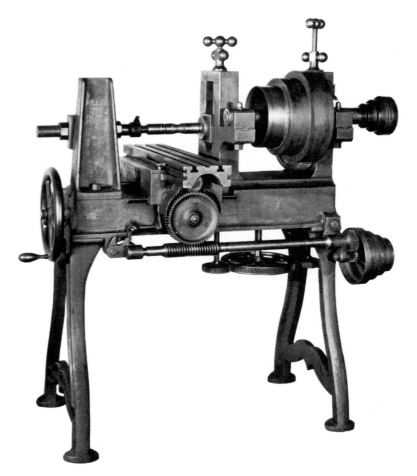

"Cut the cable!" yelled Elisha Graves Otis as he rode a crude elevator platform up a specially built frame in the Crystal Palace trade exposition in New York in 1853. A workman swung his axe and obliged the passenger. The car dropped an inch and clamped securely on ratchets installed on the girders of the shaft. Otis had proved in the most dramatic way, at the risk of his own life, that he had fortified the safety of passenger elevators. Otis, an exact contemporary of Isaac Singer, had craft experience in building and mechanics. In 1852, when he was 41, he was given the job of installing machinery in a new factory at Yonkers, New York. Elevators as lifting platforms for goods were fairly familiar. But either they were hydraulically operated and therefore very slow, or their safety record was poor. If they were hung with a counterweight over a pulley, the hemp ropes used would rot and break with little warning. At Yonkers, Otis experimented with his new safety device – a ratchet and pawl which was activated by a wagon spring as soon as there was no tension in the suspending rope. Otis left his job and set up a small business selling elevators. They were still used almost exclusively for freight, and he took few orders. At the New York exposition he decided on his demonstration, and the experiment was successful. He installed his first safety elevator, steam-driven, for the use of passengers at the New York department store of Haughwout's in 1857. The elevator traveled the five floors in less than a minute. The efforts of Otis, William Kelly, and Henry Bessemer made skyscrapers acceptable. Technical improvements in the drive of the elevator itself eventually allowed speeds of up to 1,800 feet a minute.

Isaac Merrit Singer did not invent the principle of the sewing machine, although he perfected it. His most original and influential activity was to market the domestic machine and sell it aggressively, with the inducement of deferred payments. Singer was an above-average mechanic with a couple of patents to his name when in 1851, at the age of 40, he was asked to repair a Lerow & Blodgett sewing machine. Within days he had designed an improved model which he patented and began to manufacture. The Singer machine was the first to permit continuous stitching on any part of the work being handled. By putting the eye of the needle at its point and using lockstitch, Singer had infringed the five-year-old patent of Elias B. Howe. But he agreed to pay royalties, and continued his massive schemes for manufacture and selling, using a force of 3,000 salesmen in America and an even greater sales contingent for his European agents. He made a succession of improvements in design, adapting his machine for bootmaking and garment manufacture, and retired to England in 1863 at the age of 52. In a dozen years he had created a new industry with a production surpassing the total output of all other textile machinery in the United States. Singer's sewing machine influenced the entire Western industrial world with its advanced methods of mass production, the economic potential of the ready-made clothing industry which it encouraged, and the prospect of more creative work for women.

The rate of advance of science, says Isaac Asimov, *depends a great deal on advances in the techniques of measurement*. The same is true of technology. In the 19th century no one man did more to encourage accuracy of measurement in manufacture than Joseph Whitworth. After working with Maudslay and his "Lord Chancellor," the bench micrometer which the engineer jocularly called *the final court of appeal*, Whitworth gained further experience in tool shops, then set up his own company in Manchester in 1833. He concentrated on producing machine tools for sale to industry and eventually dominated that market. At the London International Exhibition of 1862 he occupied a quarter of all the space allocated to machine tools. He designed a number of new ones, and introduced a standard thread for screws. He insisted on accuracy, and every piece machined was constantly tested against a true plane surface and meticulously measured. *What exact notion can any man have of such a size as a "bare 16th" or a "full 32nd"*? he asked rhetorically, and he introduced the system of end measurements, placing the workpiece against the master standard and noting any difference by the sense of touch. *We find in practice in the workshop that it is easier to work to the 1/10,000th of an inch from standards of end measurements than to 1/100th of an inch from lines on a two-foot rule*. In 1856 he had perfected his "millionth measuring machine," a bench micrometer graduated to measure to .000001 of an inch.

While Whitworth was completing his micrometer. William Kelly in the United States and Henry Bessemer in England were pursuing the manufacture of steel from cast pig iron. using only the carbon remaining as an impurity in the iron to burn itself out by a forced draft of air. In 1856 Bessemer presented a scientific paper on this topic. When Kelly heard of it he claimed the patent in America. but immediately went bankrupt. His interests were taken over by Bessemer. Traditionally. iron was made in a blast furnace by heating the ironstone with coal or charcoal and lime. and blasting in air. which was speedily reduced to combustible carbon monoxide. which in turn combined with carbon and oxygen in the ore. The resulting molten iron was run off to solidify in bars called pig iron. which still retained some 10 per cent of carbon and other elements. Cast iron was converted to malleable wrought iron by being reheated on the hearth of a furnace and exposed to the air. which oxidized the unwanted elements. leaving up to 0.5 per cent of carbon. Steel. which contains up to three times as much carbon as wrought iron. was made either by stopping the refining process while sufficient carbon remained in the iron. or by reheating wrought iron with charcoal. The Bessemer converter received molten iron which was subjected to an air blast percolating through vents in the lower section. consuming the silicon and carbon by oxidation. When the iron was blasted into purity a predetermined quantity of molten iron was then run in. with calculated amounts of carbon. manganese. and special ingredients such as silver. rhodium. or chromium. The blast was then re-applied for some five minutes. Bessemer process steel was avidly produced by every industrial country until the open-hearth method succeeded it. In 1870. 215.000 tons were produced in Britain alone – half the world output – and by 1900 world production was 28 million tons. Bessemer steel was cheap and strong. It demanded less fuel to make. and used varying iron ores. compensating their chemical differences by added ingredients. But Bessemer's great luck was that he marketed his process at the beginning of the railroad boom. with toolmaking. ship building and the new high buildings waiting in line for steel.

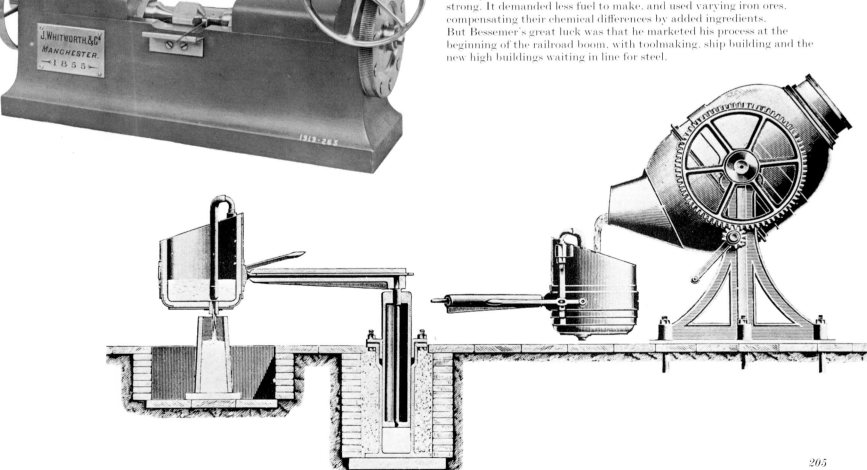

Isambard Kingdom Brunel, a visionary engineering genius never strong on finance, negotiated in the early 1850s for the construction of *a great iron ship*, designed to steam to Australia via the Cape and back via the Horn at 15 knots without refueling (there were virtually no coaling stations then).

The launching ceremony of the *Great Eastern* took place at Millwall on the side of the Thames, but the hulk refused to budge. It was eventually heaved off the slipway in January, 1858. With a length of 692 feet it was twice as long as any ship of its time, and its gross tonnage of 19,000 was never paralleled in the 19th century. The ship was of double-skinned cellular construction, having an inner and an outer hull of three-quarter-inch iron plates separated by a ribbed space of almost three feet. It was driven by two paddle wheels 58 feet in diameter and a four-bladed screw propeller of 24 feet diameter; its engines required six boilers and 72 furnaces. It had five funnels and six masts, which carried 6,500 square yards of sail. At the outset, the *Great Eastern* was damaged by a gale in Holyhead harbor. Another gale in mid-Atlantic damaged its rudder and paddles to such an extent that it rolled about helplessly for several days during which its frame showed admirable strength. When it was approaching New York it ran on a rock which stripped 80 feet of outer plates off its bottom, though owing to its cellular construction it suffered no permanent harm. After repairs it steamed home in 11 days. But confidence in it was shaken. The expected contract to employ it for passenger and mail transport on the long sea route to India did not materialize. The plain fact was that its 6,600 horsepower engines demanded 75 per cent more coal than had been anticipated. The owner companies were involved in such heavy liabilities that they sold the vessel, which had cost half a million pounds, for a third of its scrap value at £25,000. The *Great Eastern* fought back, however, and won permanent prestige. It was refitted as a cable ship and carried out the task – which it alone could have accomplished – of laying the whole length of the Atlantic telegraph cable, 2,600 miles long, and weighing 5,000 tons, in 1865. It was then exclusively employed on cablelaying until it was broken up in 1889. By that time, the St Gotthard tunnel had long been completed, cutting the rail journey from Germany to Italy by 36 hours; and the Channel Tunnel, with 2,000 yards bored out on either side, had been vetoed by the British military authorities. All that materialized from that melancholy episode was the discovery of a coal field near Dover.

But the fuel which was to succeed coal was being tapped. The drilling contractor Edwin Drake struck oil at 69½ feet near Titusville, Pennsylvania, on August 27, 1859, and within 15 years the annual output of the Pennsylvania oil field was 10 million 360-pound barrels. Previously, drilling for salt had been the only good reason why Americans should erect derricks and twist drills. Samuel Keir of Pittsburgh was angry when an expensive boring at Tarentum on the Allegheny showed brine adulterated by oil, and in desperation he skimmed the oil from the brine, bottled it, and sold it labeled as *Keir's Petroleum . . . celebrated for its wonderful curative powers as a natural remedy*. The label intrigued an entrepreneur named George H. Bissell, who first formed the Pennsylvania Rock Oil Company and then sent for full analysis a sample of oil seepage on Hibbards Farm, Titusville. Benjamin Silliman, professor of chemistry at Yale, prepared fractions of the oil by distillation, from which he extracted illuminating gas, paraffin wax, and naphtha. The latter proved to be a very suitable lamp oil, cheaper than the sperm oil and vegetable oil then used, and easier to produce than the kerosene extracted from coal and asphalt deposits. Bissell now had a keen commercial motive for conducting serious drilling, resulting in the strike at Oil Creek.

The Mont Cenis alpine railway tunnel, designed to link Paris with Turin, was started in August, 1857, using only gunpowder blasting and manual clearing and trimming. Later, new boring machines introduced by the Italian, Germain Sommeiller were worked by compressed air forced to a pressure of 90 pounds per square inch by water-powered compressors harnessing the natural heads of water found within the mountain. There were no vertical shafts leading to the exterior, since the middle of the tunnel was one mile below the mountain summit. On Christmas Day, 1870, perforator number 45 bored through the four-foot rock wall separating the two approaches, and the center lines of the two workings were found to vary by no more than four inches vertically and eight inches horizontally. With amazing speed, the tunnel was tidied, foundations prepared, and the rail track and signaling installed. The Mont Cenis tunnel, seven miles, 1,059 yards long, was officially opened in 1871. In 1872 the English Channel Tunnel Company was formed.

The first underground urban railway involved very little tunneling. The four-mile-long Metropolitan Railway in London was intended to connect Brunel's Great Western terminus at Paddington with other railway termini north of the Thames. It was mainly a "cut-and-cover" job, digging below main roads and vaulting the excavations later. It took just three years to build. After a less-than-gala unveiling, really an official inspection in which Gladstone, then Chancellor of the Exchequer, traveled with others in an open truck, the line opened on January 10, 1863. Traveling conditions were less rigorous than Gladstone endured, though steam and soot abounded. The coaches were roofed and lit with gas, *turned on so strong in the first-class carriages*, it was reported, *that newspapers might be read with ease*. Mechanical breakdowns, however, excited consumer revolt. *As I have been particularly recommended by my doctor to "select a bracing air,"* publicly complained one passenger, *and as I suffer from asthma and bronchitis, I really don't think that this detention in a sulphurous sewer near Baker Street for a whole hour is likely to improve my general health*. Unpolluting electric traction power was to come with the first genuine "tube" excavation, the City & South London Railway, opened in 1890.

207

In the same year that Edwin Drake struck oil in Pennsylvania, Etienne Lenoir of Paris designed an internal combustion engine powered by illuminating gas, and he began to manufacture these engines a year later, in 1860. A previous gas engine designed by Eugenio Barsanti and Felice Matteucci of Florence in 1853 had been working at a railway station in Florence since 1856, but Barsanti and Matteucci set up their manufacturing company in the same year as Lenoir. Historically, the first internal combustion engine had been a device of Christiaan Huygens, the Dutch astronomer and physicist who became one of the charter members of the Royal Society in 1663. He made a machine whereby a small measure of gunpowder exploded in a cylinder and raised a piston which was forced down by atmospheric pressure when the gases cooled. As in the original Newcomen atmospheric steam engine, this downward stroke provided the working action. Steam, as an expansive force, took over from the somewhat hazardous gunpowder, but Lenoir conceived the use of an explosive mixture of gas and air. The construction of his gas engine was virtually that of a double-acting steam engine with a cylinder, piston, and flywheel. The gas mixture was fired inside the cylinder by a spark from a battery and induction coil when the piston was in the middle of its stroke. The return stroke of the piston expelled the exhaust gases on one side and drew in more gas and air on the other. What was lacking was any previous compression of the gas mixture. Although Lenoir's gas engine cost more to run than a steam engine of comparable power, it became popular as a convenient, compact industrial power source.

Lenoir had entrusted the manufacture of the gas engine he designed in 1859 to a Paris engineering firm, Gautier & Cie, which retained him as consulting engineer. Lenoir continued to experiment with internal combustion engines, particularly as a means of propulsion for vehicles. His gas engine was too cumbersome for this purpose, and he examined hydrocarbons as fuel. The most convenient hydrocarbon at that time was benzene produced from the distillation of coal tar. Lenoir built a small engine running on benzene and giving one and a half horsepower at 100 revolutions per minute. In May 1862 he fitted a similar engine into a magnificent three-wheeler.

From the design standpoint the vehicle was far more functional than the shaftless horseless carriages which have been erroneously considered to be the first motor cars; it did, in fact, have a sound traditional style, evolving from the steam carriages which ran on public highways in the 1830s. This car may have burnt benzene, since the tank shown in the extant mechanical drawing seems a meager container for three hours' supply of gas, and Lenoir did accomplish a three-hour journey the following year. The problem of converting a hydrocarbon liquid by vaporization into an inflammable gas is reputed to have been solved half a dozen years later, in 1867, when a carburetor was built by Siegfried Marcus of Vienna. That carburetor seems to have been ignored at the time. So also were the petrol-engined handcart Marcus is said to have constructed in 1864 (in fact, he built it in 1870 after he had developed his carburetor), and the motor car still credited to him and dated 1875 (though it was actually built by other hands in 1888 and never exhibited until 1895). In September, 1863, Lenoir ran his machine publicly in Paris. He built a replica for Czar Alexander II of Russia in the following year, by which time he was more interested, and marginally more successful, in fitting new designs of internal combustion engines into motor boats.

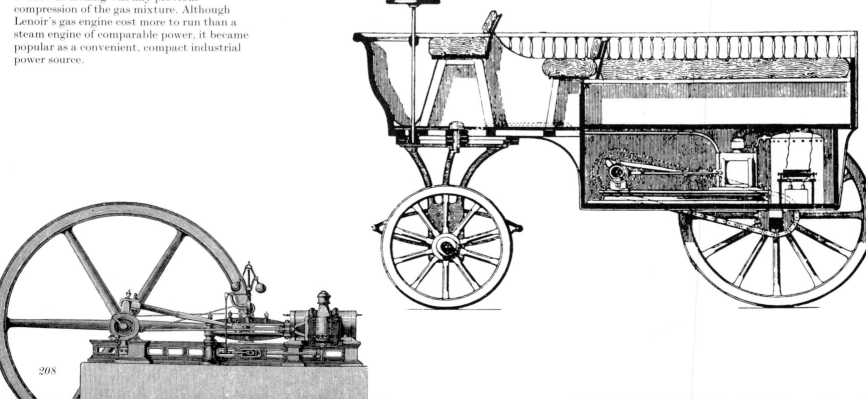

208

Nicolaus August Otto was a man of mechanical bent and no resources, who survived on the commission he earned as a commercial salesman. Reading a newspaper account of Lenoir's gas engine in 1860, when he was 28, he built one based on the scanty details he had gleaned. He tried to control the force of the explosions, which was always a shattering problem with the Lenoir engine. He failed, and as an alternative built a vertical one-cylinder engine in which the gas explosion drove the piston up and the combination of the weight of the piston and atmospheric pressure brought it down for a working stroke. This was a reversion to the engines of Huygens and Newcomen, but Otto did devise an efficient one-way clutch that would pass the piston up without impediment and transmit the action cleanly on the working stroke. Otto went into partnership with a shrewd capitalist, Eugen Langen, and from 1867 the Otto & Langen atmospheric engine was a popular production model. Otto, however, wished to improve it radically by making it much quieter and giving it more power. The only answer was internal combustion of a *compressed* mixture of gas and air, and smoother running based on a four-stroke cycle in the cylinder. This had already been outlined theoretically but was probably not known to Otto, who worked it out himself.

In the four-stroke cycle the piston draws in gas and air; it compresses the mixture, which is ignited; the expanding gases drive the power stroke; and the piston, kept to its rhythm by a steady flywheel, expels the exhaust. The Otto Silent Engine of 1876, yielding some three horsepower at 180 revolutions per minute, had been sold to its limit of 200,000 by the end of the century. Though still a single-cylinder, it was the ideological precursor of the automobile and airplane engine.

Gottlieb Daimler had come to work with Otto & Langen after previous experience at Whitworth's in Manchester, England. He was the production manager for both the 1867 and 1876 machines. Daimler wanted the firm to concentrate next on a light, portable engine which ran at high speed and burned a gas which could be vaporized from a liquid. Otto was not convinced, and Daimler left the firm, taking with him another key engineer, Wilhelm Maybach. The two set up in Cannstadt in 1882 and built light high-speed engines burning vaporized benzene. At the same time Carl Benz was working independently on a benzene-fueled internal combustion engine. The automobile was just three years away.

Kirkpatrick Macmillan, the Scottish blacksmith who adapted the hobbyhorse for mechanical propulsion, did not set the world on fire. There was no public demand for his creation; just occasionally he put together a machine for a relative. Bicycling by the treadle-and-crank principle was stillborn. However, 22 years later a Paris coachbuilder, Pierre Michaux, accepted a *vélocifère* – a modified model of the old *célérifère* – for repairs to its front wheel. His son Ernest later took the machine out for a test run. When he complained that his legs got tired as he tried to keep them off the ground, his father suggested a step on either side of the front axle for the rider to rest his feet. *Or preferably*, he continued, *fit a cranked axle to the front wheel and turn the crank with your foot*. Ernest put his father's idea into execution and produced two prototype bicycles. The following year, in 1862, the family set up as manufacturers and made 142 machines. The boneshaker had arrived. Gradually the size of the front wheel was increased until the profile reached "penny-farthing" proportions. And in 1869, when a Perraux single-cylinder steam engine was fitted to this pedal-and-belt-driven missile, it became the pioneer motorcycle. The original styling was retained, demanding iron nerve from the rider with a boiler under his saddle.

James Clerk Maxwell, whose stature in the world of science derives from his use of mathematics to deduce the molecular theory of gases and to establish electromagnetic radiation, also made valid advances in astronomy and in the study of light. A spin-off from experiments he made with the prism was a historic color photograph which he demonstrated to the Royal Institution on May 17, 1861. He had commissioned Thomas Sutton to photograph a tartan ribbon bow against a black background, using successively filters of red, green and blue between the object and the lens, and taking three separate photographs. Transparent positives on glass slides were made from the collodion negatives and were projected by lanterns simultaneously, through their appropriate filters, to make an agglomerate image in precise registration on a screen. Spectators were impressed, but the fidelity of the effect could not be judged until in 1939 techniques were available which combined the original negatives to make a colour print permanently recording the 69-year-old image.

Tis not impossible, Robert Hooke had said in 1684 after experimenting with a taut wire, *to hear a whisper a furlong's distance . . . Perhaps the nature of the thing would not make it more impossible though the furlong should be 10 times multiplied*. Electrical power was applied to telephony in 1860 by Johann Philipp Reis, a 26-year-old violinist of Friedrichsdorf, Frankfurt-am-Main, when on makeshift apparatus he sent violin notes over a short distance. He picked up the sounds through a wooden tube vibrator – the hollowed bung of a beer-barrel crowned with stretched sausage-skin – and he used his violin-case as a resonator (a non-electric loudspeaker). In the following year, on October 26, 1861, with more sophisticated gear, he sang songs which were imperfectly received 100 meters away at a demonstration before the Physical Society of Frankfurt. His vibrating diaphragm made intermittent contact with a metal point in an electrical circuit, and an electromagnet in his receiver transmitted the sound through his resonator.

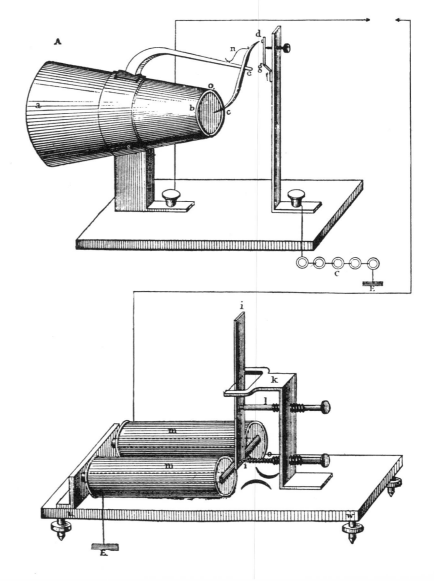

Frederick Howe, the armory mechanic whose milling machine had been modified by Root and Pratt to produce the Lincoln Miller, collaborated with another precision designer, Joseph Brown, to produce the far more sophisticated "Universal Milling Machine." This machine incorporates every basic feature of the models used today. The initial problem it was designed to surmount occurred again in small arms manufacture. Howe had transferred from the defunct Robbins & Lawrence to the Providence Tool Company in Rhode Island, which was making the new percussion lock rifle. Accurate twist drills were required to drill the holes in the detonator nipples. The helical flutes of these drills were at that time slowly filed by hand on steel rods. When demand escalated with the American Civil War, Howe sought to produce the drills mechanically. He was a friend of Joseph Brown, whose neighboring firm of Brown & Sharpe served as instrument makers and manufacturers of precision gauges. In 1862 Brown produced a universal milling machine which machined the spiral flutes of twist drills automatically and varied their pitch by the use of change wheels. When equipped with milling cutters which Brown designed, the machine also cut gears which previously had been hand-formed.

The turret lathe had been built by Stephen Fitch to do eight successive jobs before the machine had to be stopped to change tools. It was given a further push towards full automation by the work of Christopher Miner Spencer. Spencer, an effervescent genius brimming with ideas, was born in Manchester, Connecticut, in 1833 and won early appreciation from Pratt & Whitney, a young Hartford firm, which manufactured many of his patented machines. Spencer's seven-shooter repeater rifle had notably influenced the course of the American Civil War, but he was more at home with machine tools for textile manufacture. After the war, encouraged by his success in building an automatic lathe for turning sewing machine spools, he converted a Pratt & Whitney manual turret lathe to automatic operation for turning metal screws. The central principle of the conversion was what Spencer termed his brain wheels. In the perfected 1873 version these were rotating drums, on the outside of which flat strip cams could be mounted to suit any required operation. The cams engaged metal fingers which controlled the operation of the slide, the turret containing the tools, and the collet chuck. The result was a high-production automatic precision tool, conventionally called a screw lathe but in reality capable of a great variety of turning operations. Spencer took automatic action even further by making three turrets operate automatically and simultaneously on one workpiece. The principle was established, and what had begun as an automatic screw machine led to the complex specialized machine which today, within a few minutes, turns a rough metal casting into a sophisticated, precisely engineered production piece.

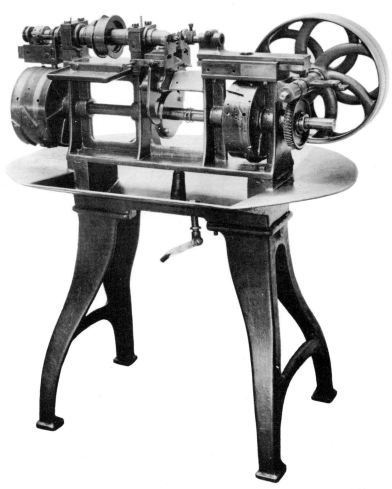

Efforts to produce a practical typewriter abounded through the 19th century, and a heavy Danish flatbed plunger-printing machine went into commercial production from Copenhagen in 1870. But by that time, Christopher Latham Sholes, a Milwaukee printer, with his associate, Carlos Glidden, was working his way through the construction of 30 successive models. Gradually he improved his basic conception of 1867, using the principles of a single character on each type bar; a horizontal platen moved one space forward after each impression by means of a weight, like a clock escapement, and an inked ribbon between the type metal and the paper. The worst problem was the clashing and tangling of adjacent typebars when one went up and the other came down. Sholes's solution was to abandon alphabetical order and position the characters in places where adjacent keys were least likely to be used consecutively. In 1872 Sholes introduced his machine to the public; and his financial partner, James Densmore, set about the organization of mass production and distribution. The Remington Small Arms Company of Ilion, New York, being comparatively idle between wars, had diversified into making agricultural implements and sewing machines. They were now persuaded to manufacture the Sholes & Glidden Type-Writer. In 1876 the machine was marketed under its new name as the Remington No.1. The Remington No.2 of 1878 had a double keyboard, introducing lower case as well as the capital letters which had previously been the sole provision.

Sales were surprisingly slow at first, but even in Society the machine had a novelty value: it was certainly not then considered inappropriate for love letters, but its strangeness tempted even first-class transatlantic passengers on the White Star line to peer over the writer's shoulder. Winston Churchill's jaunty uncle, Moreton Frewen, bought an American typewriter to make a dubious company prospectus look respectable, boarded the liner *Servia* and wrote to the bride he had left in New York: . . . *I am writing this in the saloon, and, this machine being a novelty, people will stand around to watch me write. But when I get to London and to solitude I will pour out all my heart to my darling wifie who has made my life so dear and delightful to me.* From the London residence of Lord Randolph and Lady Jennie Churchill he did pour out his heart – through a purple ink ribbon. And on Winston's 19th birthday he gave him a typewriter – which Churchill used neither for company-promoting nor for love-making but for writing war dispatches.

The technical lead to telephonic communication had to be founded in comprehension of electromagnetic induction, which Faraday had demonstrated in England and Henry had put to practical use in both America and England, presenting Morse and Wheatstone with the electric telegraph. But success depended also on a more refined understanding of the connection between the vibrating frequencies of sound and their effect on electric current. The telephone pioneer Alexander Graham Bell was the grandson of a Scottish actor who taught elocution in London, and the son of a specialist in teaching the deaf and dumb. He himself taught speech, studied anatomy and physiology, and finally qualified as a doctor. Threatened at 23 with tuberculosis, from which two brothers had died, he emigrated with his family to Ontario in 1870 and with his father again taught speech to the deaf. But soon he was in Boston, where he was appointed professor of vocal physiology.

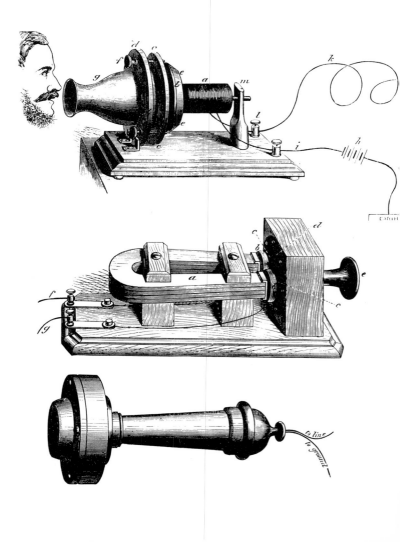

212

In his keenness to develop an efficient hearing-aid Bell explored every connection between electricity and acoustics. While working on a method of sending multiple messages simultaneously along a telegraph line by sending the signals at different pitches, he learned that different tones varied the strength of an electric current in a wire. At the Massachusetts Institute of Technology he examined a machine which he recognized as mechanizing the human ear, a vibrating membrane producing a reproducible physical effect. Consulting Joseph Henry, then nearing his 80th year, Bell constructed a microphone in which the membrane would vary an electric current, and a receiver which would reconstitute the variations into audible frequencies. On March 10, 1876, an accident occurred when what was virtually the "office intercom" had been left activated while equipment was being set up. Bell spilled battery acid on his trousers and called *Mr Watson, come here – I want you* – and the first telephone message was passed and received.

Bell used his royalties and prize money to set up schools and research for the deaf, improve Edison's phonograph, transmit sound by light in a pioneer electronic advance, and greatly forward many other branches of science, particularly aeronautics. When he was 72 years old his hydrofoil HD-4 took the world water speed record.

Thomas Alva Edison, just three weeks younger than Bell, had played tit-for-tat by swiftly improving the telephone in 1877 with a carbon-powder microphone/mouthpiece. In the same year he produced his phonograph, intending it to be used as a speech recording or dictation machine for commercial offices; an idea derived from an unpursued scheme for an "answer-phone" system (though there was then no telephone network) for filing telegraph messages. He built a mouthpiece holding a diaphragm, in the center of which a stylus was held in contact with the grooved tinfoil surface of a hand-wound screw-progressing cylinder. The vibration of speech into the mouthpiece while the cylinder was being wound caused the stylus to transmit a characteristic pattern. After winding back the cylinder, the message could be taken off at any later time by a stylus on the opposite side of the cylinder, riding in the grooves and reproducing the sound by a diaphragm almost identical to that in the mouthpiece.

Clerks were cheaper than office equipment, and Edison's phonograph fell on commercially stony ground, bringing in a little money only by being rented to traveling showmen and downtown amusement parlors as the latest freak. (The showmen were sure of their money, since nobody could hear a thing without a tube and earpiece, supplied after receipt of the dime.) In 1886, after four years' research financed by Alexander Graham Bell, Alexander's cousin, Chichester Bell, and Charles Sumner Tainter brought out their graphophone, which had a wax-coated recording cylinder. The challenge of a rival spurred the busy Edison to improve his own recording cylinders and to market them aggressively as a source of entertainment – not as blanks for home recording, but already holding professional recordings. Edison's phonograph, clockspring-driven from 1895, was thus set to become not only his personal favorite, but by far the most financially rewarding of all his 1,093 patented inventions.

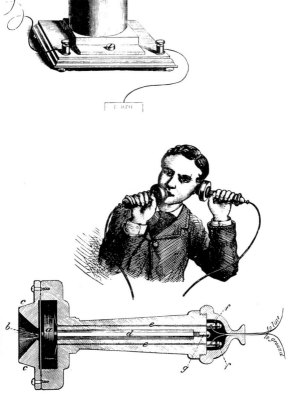

Zénobe Théophile Gramme, a Belgian living in Paris, produced the first practical dynamo to generate direct electric current continuously in 1871. He improved it considerably in 1878. This was the consummation of the idea of the electric motor and generator which Faraday had sketched as early as 1821 and 1831, demonstrating under laboratory conditions that electricity could make a pivoted magnet revolve, and that a disc being turned in a magnetic field would induce an electric current. Forty years passed before technology could catch up with theory, when Gramme produced a ring armature which did not overheat in action and could be incorporated in a steam-driven dynamo for continuous use. This was seized on for illuminating lighthouses and factories (and even the Gare du Nord), and for such low-powered processes as electroplating. In 1873 the Gramme factory in Paris was itself wired for power by motors running from the Gramme dynamo – the first instance of electricity being used to supply mechanical power on an industrial scale. Eventually the system proved unsatisfactory because of the constant need to service the commutators and brush gear, and Gramme introduced single-phase ring-wound alternators. Meanwhile, Gramme had shown that his generators could be used in reverse, a dynamo being set up to provide electricity to drive a second ''dynamo'' as a motor, though not efficiently.

The perfection of his dynamo in 1878 enlarged the scope of electric motors for industrial use, though they were still confined to direct current. More portable sources of electric power were still practical: specially designed batteries driving a Gramme electric motor were to power the airship *La France* on the first truly dirigible aerostatic flight in 1884. Apart from the industrial use of electricity, it was in growing demand for illumination. Until the 1870s public electric lighting was achieved by arc lamps. These needed constant adjustment of the spaced electrodes and were inefficient. In 1876 Paul Jabochkoff, a Russian military engineer who had settled in Paris, introduced his ''candles'' – arc lights which needed no adjustment but which required high-voltage current. Gramme produced a generator to serve them. But the system still proved uneconomic, and it was clear that the future of illumination lay in the incandescent filament lamp.

This required a durable filament which would not melt or burn away. It demanded an efficient vacuum inside the lamp, which was not practicable until Hermann Sprengel's mercury vacuum pump was introduced in 1865. Sir William Crookes made a public success of this when he evacuated glass globes for his radiometer in 1875. Joseph Swan, a Newcastle chemist who had been experimenting with incandescent filament lamps since 1848, became encouraged by Crookes' discovery. On December 18, 1878, Swan exhibited a carbon filament vacuum lamp to his city's Chemical Society, and began urgent work to get it into production. He used for the filament a carbonized thread of mercerized cotton – cotton treated with caustic soda. In the same year Thomas Edison, looking around for an invention he could profitably exploit, began working on the development of the incandescent filament lamp. On October 21, 1879, he registered success with an evacuated lamp containing carbonized slivers of bamboo. Previously he had worked with filaments of platinum, and he still retained platinum as the conductor passing through the base of the lamp to the source of the current, because it was the only conductor known to have the same coefficient of expansion as glass. Swan had taken out no patent. Edison character-istically patented every notion. Swan's manufacturing enterprise was blocked by Edison, but Swan introduced further refinements which he patented, and barred Edison. After a tentative law suit they combined to form the Edison & Swan United Electric Light Company in 1883.

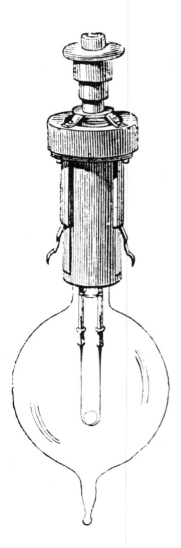

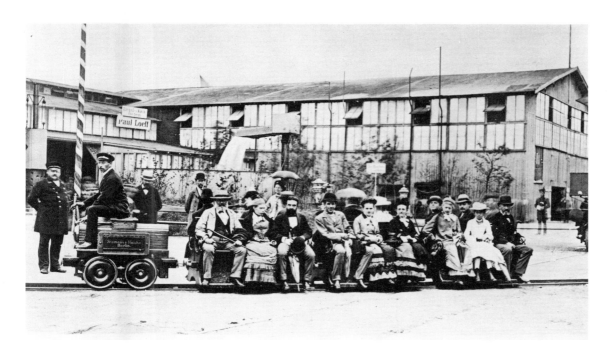

It was typical of George Stephenson, often referred to as "the father of the steam locomotive" (although he did not invent it), that in his old age he should have predicted that a time would come when electricity would be the great motive power of the world.

The first really practical electric locomotive which could haul a load of passengers around a line was the work of Werner von Siemens, who demonstrated it at the Berlin Trade Exhibition in 1879. The tiny three-horsepower locomotive took power from a stationary generator by means of a center live conductor rail. The same firm, Siemens and Halske, followed their Berlin demonstration by equipping a short section of line between the railway station and the military academy in Lichterfelde in 1881. This is generally accepted as the first electric railway in the world to provide a regular commercial passenger service.

Already Edison had been busy in New York with propaganda and intricate organization to get his lamps into the homes of the people. Using the strength of his own wealth and prestige, he acquired the finances to wire houses, install domestic generators, lay underground cables to premises which would accept public distribution, and above all to generate the power necessary for this massive projected supply. In 1882 he opened the Pearl Street power station in New York City, the first of all public generating plants. He saw this expensive installation system as the overhead of a "razorblade industry" concerned primarily with furthering the sale of electric lamps. Even by 1885 he was selling 200,000 lamps a year at a price which would be high today: $2.50.

In a book on optics written about 130 AD, the astronomer Ptolemy discussed persistence of vision – the fact that the retina of the eye continues to record an impression for a moment after the image is no longer there. Throughout the centuries it was used for many a pleasant parlor illusion, and it is the property of the human eye which makes cinema and television work. In the mid-19th century the illusion was most skillfully exploited in the praxinoscope – another typical Victorian hybrid term based on the Greek words for *action* and *observe*. The praxinoscope was painstakingly described in the literature of the time as *a scientific toy in which a series of pictures representing consecutive positions of a moving body are arranged along the inner circle of a cylindrical or polygonal box open at the top, and having in the middle a corresponding series of mirrors in which the pictures are reflected; when the box is rapidly revolved, the successive reflexions blend and produce the impression of an actually moving object.*

By 1882 the praxinoscope had been adapted in Paris so that its images were projected on a screen for exhibition to a large audience.
In that year Etienne Jules Marey, a professor of medicine with an enthusiasm for recording movement, began a belated study of photography and constructed his *fusil photographique* (photographic gun), with which he could take 12 consecutive pictures in one second, spaced on the circumference of a revolving circular glass photographic plate. This, though unacknowledged at the time, was pioneer cinematography.

Eadweard Muybridge was born in 1830 in Kingston-on-Thames and was christened Edward James Muggeridge. He emigrated to the United States in 1852, and by 1856 was in business in San Francisco as E.J. Muygridge, bookseller and print publisher. In 1860, having finalized his name in a Saxon form as Eadweard Muybridge, he began an overland journey to New York. He was thrown from the coach when the six horses bolted in Texas, and with the $2,500 in damages he extracted from the stage company, he set up in San Francisco as a professional photographer. He published a remarkable series of landscapes of the Yosemite valley which were followed by many unique documentary records of California. In 1872 the former governor of the state, Leland Stanford, asked him to photograph Stanford's fast horse Occident trotting at full speed. Not until 1877 did Muybridge publish a photograph of the horse Occident, taken in less than 1/1000th of a second, while the horse was trotting at the rate of 36 feet per second. This print aroused professional controversy, and Governor Stanford commissioned Muybridge to *have a series of views taken to show the step at all its stages, so as to settle the controversy among horsemen about the question whether a fast trotter ever has all its feet in the air at once.* Muybridge eventually rigged up 12 (and later 24) cameras activated by electromagnets fired by trip wires 21 inches apart. The international success of this series of 'instantaneous photographs'' inspired Etienne Jules Marey of Paris to develop his photographic gun. Marey's subsequent pictures of the animated zoology of birds in flight, and of smoke swirling in a wind tunnel, were closely studied by the Wright brothers and by later aerodynamic designers. Muybridge mounted drawings made from his photographs of the horse on the circumference of a lantern-slide disc, revolving and projecting images onto a screen by what he called his zoopraxiscope. He went on to photograph humans in vigorous motion, running, wrestling, and jumping. On January 16, 1880, Muybridge gave a public show in San Francisco of his screened motion pictures. *Mr. Muybridge,* said the *Daily Alta California, has laid the foundation of a new method of entertaining the people.* Edison's peepshow kinetoscope was still 11 years in the future, and the *cinématographe* of the Lumière brothers over 15 years away. In May, 1888, Muybridge met Edison and discussed a combination of Muybridge's moving pictures with Edison's phonograph. But ''talking pictures'' were 40 years in the future.

In 1884 Paul Nipkow of Germany took out a patent for the transmission and reception of signals passed through a mechanical scanner and deriving from the rotation of polarized light in a magnetic field. In his transmitter he used a resistance made of selenium, an element which passes current in proportion to the amount of light falling on it. Sadly, the reaction of selenium to light was too slow to pass the signals from the scanner. However, John Logie Baird used the Nipkow discs during his early experiments in television in 1925.

The internal combustion engine was still one year away from successful development when controlled electricity was conducted back to the skies in 1884. Prompted by the search for a suitably light source of power for aircraft, the French engineers Charles Renard and A.C.Krebs used electric storage batteries to power the streamlined airship *La France*. The airship was longer than Giffard's craft but required only three-quarters of the volume of gas to carry its 4,000-pound gross weight. Its Gramme nine-horsepower electric motor derived current from comparatively light chromium chloride storage batteries which Renard himself had developed. The power unit weighed only 210 pounds, a remarkable ratio of some two and a half pounds per horsepower. A broad tractor propeller, 20 feet in diameter, fixed at the front of a 108-foot-long bamboo car, pulled the airship at up to 14½ miles per hour, and allowed a circular course and the return to base.

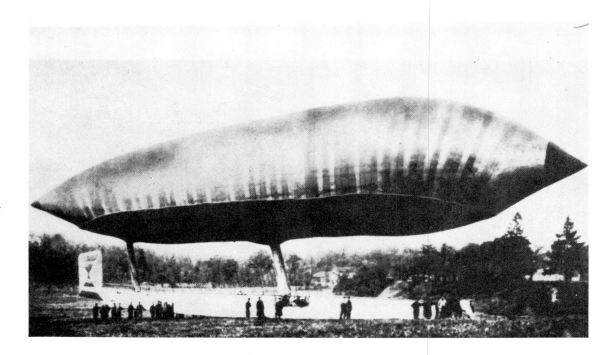

The success of *La France* was followed by the use of electric motors in undersea propulsion. Bushnell's *Turtle* had been followed by *Nautilus* which Robert Fulton painstakingly canvassed to three governments before he committed himself to surface steamboats. The Confederate forces used hand-screwed submersibles to attack the Federal fleet in 1864, by which time the French had built a 420-ton self-propelled submarine driven by compressed air. This did not provide effective range, and steam was briefly used for under-water propulsion in a submarine built by the Swedish gunmaker Nordenfelt to the design of the Rev. G.W.Garrett, a Liverpool clergyman. But the introduction of an electrically driven submarine in 1885 by the French engineer Claude Goubet marked the way ahead. Not for the first time, the constructive science fiction of Jules Verne showed its influence: in 1870 Captain Nemo had been reported traveling *20,000 Leagues Under the Sea* in his imaginary *Nautilus*, and Goubet had followed up the lead.

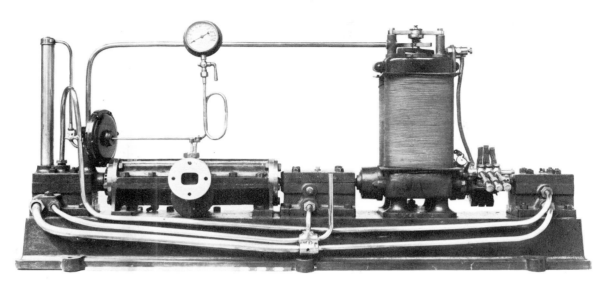

Charles Algernon Parsons was a rare combination of the handyman and mechanical theorist. He used these gifts to convert the steam engine from comparatively inefficient reciprocal action to the logical simplicity of rotating a shaft directly. In 1884 he completed his first practical steam turbine. There was a mounting need for a high rate of rotation to drive powerful generators of electricity, and advances in metallurgy had produced steel alloys strong enough to constitute working parts that would stand the strain, for Parsons' first turbine ran at 18,000 revolutions a minute. It had 14 rows of circular blades capable of revolving on a shaft inside a cylinder. Fixed blades were mounted to the inner circumference of the cylinder. High-pressure steam was projected through the cylinder and, guided by fixed blades, turned the rotor blades. The blades increased in diameter farther from the source of the steam, giving the steam its greatest efficiency. Modifications and improvements advanced the power of the turbine from a capacity of 75 kilowatts in the first dynamo installed to a typical output of 60,000 kilowatts generated in modern electricity supply.

By 1894, after legal troubles when Parsons temporarily lost the control of his patents, he applied himself to steam turbine engines for ships. For some time he faced the problem of gearing down the fast revolutions of his shaft to drive propellers at a practical speed. In 1897 he had built *Turbinia*, a 100-foot craft, displacing 44 tons, in which he planned to demonstrate the marine steam turbine. The vessel had three cylinders, driving three screws, respectively at high, intermediate, and low pressure. The propellers themselves were larger than those driven by a reciprocating engine. At the Naval Review held for Queen Victoria's Diamond Jubilee in 1897 *Turbinia*, developing 2,000 horsepower, made a surprise appearance. It darted through the Fleet at a speed of 34½ knots – nearly 40 land miles per hour. The boat could not be caught by the destroyers sent to intercept it, with their maximum speed of 28 knots. The Admiralty, shaken by the demonstration, ordered two destroyers with steam turbines, and *Viper* produced 36½ knots in 1900. Arrangements were made to install the new system through-out the Royal Navy: *Hood*, commissioned in 1917, had four geared turbines developing 150,000 horsepower. Commercial navigation took advantage of the new developments. By 1904 the Cunard Line had installed steam turbines first in *Carmania*, then in the twins *Mauretania* and *Lusitania*.

Michaux's front-axle pedal to the old hobby-horse had made a velocipede out of a *vélocifère*. Since the machines were driven by direct action, not gearing, they could achieve a higher top speed only by increasing the size of the front wheel, and therefore the distance covered in one revolution of the pedals. The racing cyclist H.L.Cortis managed a wheel of 62 inches. Braking while travelling at high speeds inevitably led to spills. The safety bicycle was designed to justify its title, and it owed its existence to the development of the driving chain. James Slater initiated it for textile machinery in 1864, but it was greatly improved by his successor in the Salford factory he ran – the Swiss-born Hans Renold, who introduced the bush roller chain in 1880. This late date is an indication of the rate at which precision and strength could reliably be built into the moving parts of a machine. The driving chain, geared to give the wheel more revolutions than the pedal but still operating on the front axle, was incorporated in the bicycle after 1878, and the diameter of driving wheels tended to drop. Experimental rear-wheel chain-drive bicyclettes began to be made. The new style was clinched by the Rover Safety, built by John Kemp Starley of Coventry and marketed from 1885. Its design eventually had wheels of equal size, and the transfer of the rider's weight to the rear of the machine gave better balance, safer braking, and finer steering. Wire-spoked wheels and tubular frames were already standard, and Dunlop's pneumatic rubber tire was introduced in 1888.

In Cannstadt, Gottlieb Daimler had been continuing his quest for a light, high-speed internal combustion engine. In 1883 he produced a vertical single-cylinder four-stroke gas engine with a speed so high that he could use the heat it generated to ignite the explosive mixture of gas and air. By 1885 he had converted this for petroleum-based fuel – at first, benzene – which was fed into a float chamber through which air bubbled from below. Ignition was effected by building into the cylinder a small platinum tube which was heated to red-hot incandescence. Delighted with his engine, which produced half horsepower at 750 rpm, Daimler determined to make it mobile. He fitted it onto a specially built bicycle held upright by outrigger wheels. On November 10, 1885, he mounted his son Paul on it and sent him off on a three-kilometer test run. The following year, Daimler built a one and a half horsepower engine running at 900 revs and installed it in the back of a shaftless carriage. In 1889 he rationalized the four-stroke system by building a V-type two-cylinder petrol engine using one crankshaft.

Carl Benz, working in Mannheim independently of Daimler, had not shared Daimler's enthusiasm for high-speed engines. His first internal combustion engine, a horizontal, single-cylinder, water-cooled four-stroke also running on benzene, gave three-quarter horsepower at 250 rpm. He used a surface carburetor with saturated wicks protruding from it, across which hot air was blown. His ignition system – battery, coil, and sparkplug – was in essence the one adopted for all later automobiles. This engine was built in 1885. Benz always aimed at fitting his engine into a vehicle, and he constructed one for his purpose – planned, not converted from a carriage. It was a three-wheeler in which a flat belt drive, which could be shifted to a neutral set of pulleys for revving up, ended in a chain drive to the back wheels. This automobile ran in 1885, and on January 29, 1886, Benz registered his Patent Motor Car. Commercially, this pioneer automobile manufacturer was not immediately successful. But he continued to improve his three-wheeler design and in 1888 sold a French concession for the machine to Emile Roger. Daimler, who was always more interested in engines than in cars, sold his French rights to Emile Levassor, who put the engine at the front of the Panhard-Levassor. He dropped Daimler's belt drive, with its smooth but slipping variable speed transmission, and introduced the lathe-type clash gear which has jangled many ear drums over the following century: *C'est brutal mais ca marche* was his grim comment. *It's rough, but it works*. Roger developed the Benz into a four-wheeler and demonstrated it in 1893. In that year Daimler first used the float-feed carburetor of his partner Wilhelm Maybach. In this, the suction of the cylinder inlet drew a spray of petrol from the constant level of the chamber to be injected into the air intake from the choke. This system soon became universal.

The United States did not seriously begin production of internal combustion engines for vehicles or for stationary power supply until the original patents for the Otto gas engine ran out in 1890. The first American-built gasoline automobile was made by Charles and Frank Duryea in 1893. The automobile industry was in fact forestalled by mechanization in American agriculture. In 1892 a gasoline-powered tractor was constructed by John Froelich of Iowa, who cannibalized a gas engine and a steam engine, adding his own innovations. A purpose-built tractor was made by the Case Threshing Machine Company of Racine, Wisconsin, in the same year; and in 1893 Charles W. Hart and Charles H. Parr manufactured 15 pioneer agricultural haulage engines, which were officially termed tractors in 1902.

On March 4, 1880, the *New York Daily Graphic* printed a half-tone picture showing a section of New York's Shantytown photographed by Henry J. Newton and technically processed by Stephen H. Horgan. This revolutionized newspaper illustration, allowing "instant" news photographs to be reproduced immediately and faithfully. The half-tone process had been initiated in Stockholm in 1871 by Carl Carleman, an engraver with a self-destructive enthusiasm for reproducing photographs in print without having them previously engraved. He photographed the original print through a screen which was crisscrossed with lines; this produced an image in which the light and shade of the original were broken into lines of varying density. Stephen Horgan in America adapted this, substituting a screen which broke the image into dots of different intensity, the heavier dots reproducing the dark areas and the smaller dots the light. Blocks made by this process could be printed at the high speed at which newspapers were produced, so that illustrations could be printed simultaneously with and alongside the text to which they related instead of separately, as in books.

In July, 1886, the *New York Tribune* adopted Linotype composing, a highly economical method of casting type for newspapers. Previously, fast typesetting had required three operators for each unit of copy passed through: one to read the copy and assemble it in type, one to keep the magazine of type characters stocked, and one to justify the lines, spacing them decently and dealing with overmatter. The Linotype needed one operator, and he could do the whole job in a quarter of the time previously required. It derived incestuously from the recently established typewriter – which was originally conceived as a miniature printing press – and it was not at first intended for large-scale printing. James Clephane, a stenographer who became a Washington lawyer, wanted more copies of law reports and his own deathless prose than the Sholes typewriter (later the Remington No.1) could provide. He approached Ottmar Mergenthaler, an émigré watchmaker from Württemberg, to tackle the task. Mergenthaler eventually produced a machine with a typewriter keyboard which released matrices, or master reliefs of letters, which were pneumatically blown to an assembly point and justified in lines, enclosed in papier mâché molds. Hot metal was poured into the molds to form slugs of type which were re-melted after use, and the matrices were returned to the magazine. Essential to the success of the process was the well-timed development of J.G.Benton's new punch-cutting machine, which ensured an adequate supply of accurate matrices. "Linotype" typesetting dominated newspaper production until the late 20th century advent of computer-aided filmsetting, and its introduction encountered appreciably less opposition.

222

On May 16, 1888, Emile Berliner, a 37-year-old Hamburg-born acoustic engineer with the Bell Telephone Company, demonstrated to the Franklin Institute of Philadelphia what he called a Gramophone. Within a month Edison, responding not to Berliner but to the menace of the Bell-Tainter Graphophone, publicly launched his new and improved Phonograph perfected as a business dictation machine. (The words *phonograph*, *graphophone*, and *gramophone* were all arbitrary inventions, linking the Greek words for *sound* and *write*, or *record*.) The characteristics of Berliner's gramophone were that it reproduced sound from disks, not cylinders, and from a very early stage of its development it was clear that it would be used not for business dictation or home recording but for disseminating entertainment already recorded by professionals. The gramophone was there to *play*.

The disk-type record had evolved as the most practical shape for Berliner's method of reproduction. Originally he had worked with a cylinder – of larger diameter than Edison's – on which the stylus made a lateral cut (the alternative to "hill and dale" indentation). The indented surface could not be removed from the cylinder for fixing as a master copy unless it had first been a separate thin metal plate wrapped around the cylinder. Wrapping a delicate plate round a drum, removing it, and straightening it flat before processing, was far too destructive.

So Berliner adopted the disk, despite the problem (which had frightened off Edison, Bell and Tainter) of countering the loss of quality caused by the increase of linear speed of a stylus in a groove as it approaches the center of a circle. Berliner used a disk of polished zinc covered with beeswax which was cut by the stylus to expose the metal, the path being later etched with acid. Except for its adoption as a toy in Germany, his gramophone, then playing seven-inch records pressed from hard rubber, did not enter serious production until 1893 or receive adequate marketing until 1896. And it was in 1896 that total war broke out between the commercial interests exploiting the phonograph, the graphophone, and the gramophone. There were struggles over patents on the instruments and recording processes as well as the foreseen battles involving the lucrative end products: disks and cylinders. The Columbia Phonograph Company (which, confusingly, sold the graphophone) diversified into disks in 1902 and abandoned cylinders in 1910. Edison Bell retained cylinders for three years more. (This firm was concerned with entertainment: Edison had a separate organization to sell business dictating machines using cylinders.) The Gramophone Company had introduced both 10-inch and 12-inch shellac disks by 1903, and in the next year the first double-sided record was presented.

Wrought iron is "worked" iron, originally hammered when red hot to beat out the slag from the bloom. Its production greatly increased between 1830 and 1850 after a brain drain from South Wales had introduced new manufacturing techniques to France, Belgium, Germany, and Sweden. Its lightness and strength were progressively utilized for building, typically in the dome of the British Museum reading room, built 1854–57. Gustave Eiffel chose the rigidity of wrought iron rather than steel for the striking tower – 300 meters high and for long the tallest structure in the world – which he built in Paris to mark the centenary of the French Revolution in 1889, celebrated by the great Paris Universal Exposition in that year. The 7,300-ton wrought-iron structure, standing on a base of reinforced concrete, is principally supported by four square frames rising in a hyperbolic curve and intricately trussed for mutual bracing. It was stipulated that construction for the exhibition should be entirely a French affair, but no native company undertook to design or build the elevators which had to follow the difficult lower curve of the legs to carry passengers 116 meters to the second balcony: these were finally installed by the Otis Elevator Company.

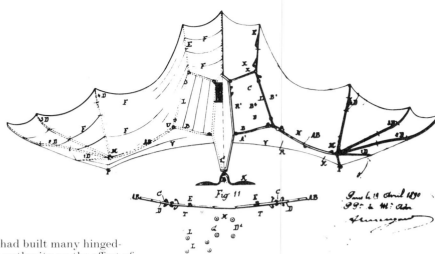

Gustave Eiffel, who had built many hinged-arch bridges, was an authority on the effect of wind on laced structures. He used all his aerodynamic knowledge to shape and support the tower. During the maximum force it has had to withstand, a gale raging at 93 miles an hour, its sway was only 12 centimeters. Toward the end of his long life Eiffel applied his specialization in aerodynamics to problems of aeronautics and in particular the perfection of the airscrew.

While Eiffel was building his tower, his compatriot Clément Ader was working on a steam engine for installation in an aircraft. His steam-powered *Eole*, the first powered airplane to take off from level ground, traveled through the air for 50 meters at a height of 20 centimeters on October 9, 1890, at the Château Pereire, Armainvilliers, Gretz, France. Ader, then aged 49, had made an early fortune from developing telephone equipment and spent it on aeronautics, beginning with a study of bird flight. His *Eole*, developed between 1882 and 1889, had wings with a span of 14 meters and an area of 28 square meters shaped like those of a bat. The airplane had a complicated provision for the wings' adjustment during flight, but while airborne for two seconds Ader had no time to operate them. Power came from Ader's very light four-cylinder steam engine, developing 18–20 horsepower and driving a four-bladed tractor propeller of three-and-a-half meters diameter. The airplane had no elevator and only vestigial controls, negated because the pilot was sitting blind behind its tall boiler. It was incapable of sustained or controlled flight, though Ader later alleged he had flown a further 100 meters in 1891. In 1892 Ader was commissioned by the French War Ministry to build a new airplane: it was the first Government subsidy ever granted for aeronautics. Financed by this contract, he constructed *Avion III*. This was another simulated bat-wing construction, with variable sweep-back wings intended to shift the center of pressure during flight, operated by two 20-horsepower steam engines. Ader falsely claimed to have flown *Avion III* for 300 meters in 1897, and even published a fake picture of his machine in the air. Though a constant propagandist, Ader, working in isolation, had little influence on the progress of aeronautics.

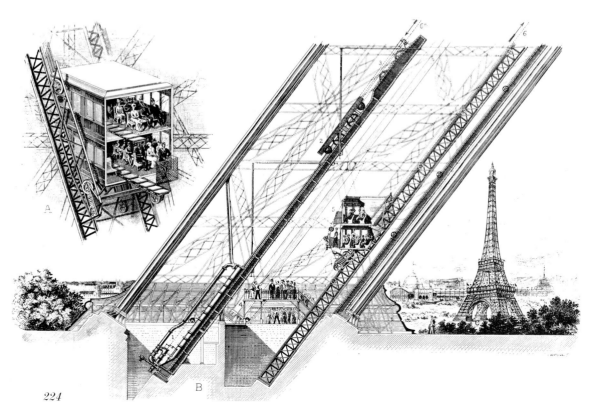

Lawrence Hargrave, an Englishman who emigrated to Australia in 1866 at the age of 16, was a trained engineer with an enthusiasm for aerodynamics. After experiments with wing-flapping ornithopters, he concentrated on the soaring and gliding of birds and began to design kites. Since he could not get sufficient stability in a monoplane wing, he put shorter wings one above the other to achieve the same lift and braced the biplane wings with vertical surfaces. He thus produced what is generally known as the box kite, which he always called the cellular kite. In 1894 he concentrated on man-carrying kites and on November 12 he raised himself 16 feet with a train of four cellular kites in a 21 mph wind. Hargrave's abiding aim was to use cellular construction to build a powered aircraft to achieve manned flight. He was a skilled engineer, and his three-cylinder compressed-air unit to drive model airplanes, made in 1884, was the first rotary engine applied to aeronautics. In 1902 he turned to steam and in 1906 to a petrol engine, driving archaic flappers instead of a propeller. The machine did not fly. Disappointed and impoverished, Hargrave gave up aeronautical research, having designed 26 unsuccessful engines. But his box-body configuration was the form adopted by many future designers.

Hargrave, working 10,000 miles away from the main centers of research, had been sustained in his enthusiasm by a set of photographs he had seen in a magazine showing Otto Lilienthal soaring in a glider. Lilienthal, a Prussian engineer, began by building a flapping ornithopter in 1869 when he was 21, but changed to fixed-wing gliders in order to master controlled stability in the irregularities of the wind. His publicized achievement was always in hang gliding – supporting himself by his arms and elbows and letting his body sway to change the trim of the machine. But his objective was to fly, not soar. In his *Bird Flight as the Basis of Aviation* he re-asserted the screwing action of the outer primary feathers by which the bird pulls itself forward, and he tried to duplicate this with artificial feathers powered as propellers by a two-horsepower carbonic acid-compressed gas motor.
He contrived a hinged tail plane to be used as a limited elevator by harness worked by his head. While testing this harness in flight he stalled, sideslipped, and crashed to death on August 9, 1896. Lilienthal greatly influenced the Wright brothers, and it was his description of a nose dive encountered after stalling that convinced them to set their elevators forward of the front wings.

Muybridge's multiple-camera action photographs of 1878 inspired Marey to make his "photographic gun," which shot 12 consecutive pictures onto the circumference of a glass plate. By 1888 Marey had exploited the new photographic paper to shoot action continuously onto a spool of paper at the rate of 12 pictures a second. George Eastman had now introduced photographic celluloid film, and Edison, with his assistant William Dickson, devised the sprocket holes to move the frames forward. Edison ran a short loop of film around a lantern in his Kinetoscope Parlor on Broadway, New York, from 1894. The subjects were typically a strong man flexing his muscles or an actress dancing. But this was not the public cinema. It was a private peep show, viewed through a slit and not projected on a screen. The Kinetoscope was known to Louis and Auguste Lumière, two brothers then in their early 30s, living near Lyon where they ran a factory making photographic materials. The Lumières devised a satisfactory method of taking intermittent pictures, stopping the movement of the film precisely when the camera shutter was opened, fixing a frequency of 16 frames a second, and above all using the same chamber, or camera, to photograph, develop, and project the moving pictures. This ensured precise registry, with the pictures always steady. On the evening of December 28, 1895, in the Salon Indien of the Grand Café in Paris, the Lumières gave a projected cinematograph screened show of a train entering a station and workers leaving the Lumière factory. People had paid to see the show. They were to line up by the thousands to pay on following nights. The Lumières, being already in a position to manufacture or license their cinematograph equipment, were the pioneers of the commercial cinema. Later they developed stereoscopic "movies."

The vacuum pump which made possible practical incandescent filament illumination within "electric bulbs" also allowed the better evacuation of the Geissler tubes in which physicists studied cathode rays – radiation from the negative electrode when high voltage passed between electrodes in an evacuated tube. In 1875 this improved vacuum cathode-ray tube was introduced by William Crookes. In 1895 Wilhelm Konrad Röntgen, professor of physics at Würzburg, experimenting with a "Crookes tube," noticed that rays from the tube were making a fluorescent surface glow even though the tube was blacked out from the glowing surface by a dark, opaque cover. He deduced that rays of an unknown nature, which he therefore called X-rays, were being radiated, and he began further experiments. He speedily recognized their resistance to the rays of metal, bone, and other dense material. The first X-ray photograph he took was of the right hand of his wife Bertha, showing her wedding ring.
The medical implications of X-rays were exploited with astonishing speed: a bullet was located in a patient's leg in the United States four days after Röntgen's first public lecture on the subject in January, 1896. (The secondary dangers of excessive exposure were not discovered for some time.) The theoretical consequences were even more important, with Becquerel's discovery two months later of radioactivity and the subsequent electron theory of the structure of the atom.

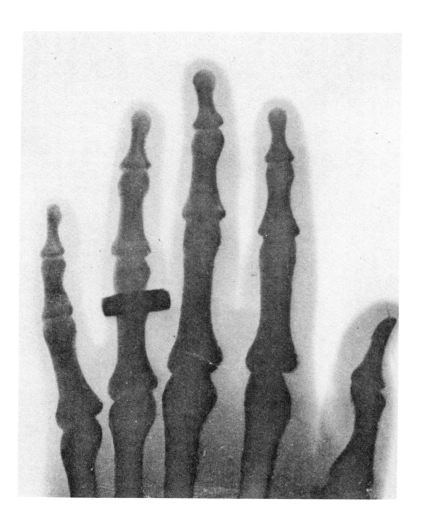

In 1894 Guglielmo Marconi, a 20-year-old student in Bologna, was investigating the discovery made eight years previously by Heinrich Hertz that electromagnetic waves (then called Hertzian waves, now radio waves) could be generated and received across space. He devised a coil and spark gap generator and a "coherer" which conducted a measurable surge of electric current when radio waves hit it. He developed the apparatus so that a key pressed at one end of a room would sound a buzzer at the other end. Failing to get financial support from the Italian government, he came to Great Britain (his mother was Irish and he spoke English) and was sponsored by the Post Office and War Office. He began to exploit "wireless transmission." By 1897 he was sending messages over a distance of nine miles, and by 1899 across the English Channel. His spectacular triumph was the justification of his belief that radio waves would follow the curvature of the earth, which would be proven if a message could be received across the Atlantic. The signal was transmitted from Poldhu in Cornwall, and Marconi sat at a receiver in St. Johns, Newfoundland. *It was shortly after midday on December 12, 1901, that I placed a single earphone to my ear and started listening. I was at last on the point of putting the correctness of all my theories to the test . . . Suddenly about half past twelve, there sounded the sharp click of the tapper as it hit the coherer, showing that something was coming, and I listened intently. Unmistakeably the three sharp clicks corresponding to three dots sounded in my ear.*

In 1906 Reginald Fessenden, a Canadian working in the United States, successfully modulated a continuous signal with variations corresponding to the frequencies of sound waves, and began the radio transmission of music and speech. Magnetic recording of speech had already been achieved in 1898 by the Telegraphone of Valdemar Poulsen, a 29-year-old engineer with the Copenhagen Telephone Company. This instrument, transmitting through a microphone, magnetized a fast-running steel piano wire (passing at 213 centimeters a second) so that variations in the recorded magnetic field corresponded to the input signal. This pioneer magnetic recorder was also extended – in theory and patent rights, but not in practice – to include the use of steel ribbon or paper tape coated with metallic powder for magnetized tape recording. Poulsen found no financial backing in Europe, and briefly exploited magnetized wire dictating machines and telephone recorders in the United States. But wire- and tape-recording was virtually abandoned until the resuscitation of the system in the 1930s. Poulsen went on to do valuable work in radio communications.

A monorail system introduced by Lartique after experiments in Belgium and France was the pioneer one-track railroad built for the 9¼-mile Listowel & Ballybunion Railway in Ireland, inaugurated in 1888. The weight of the train was taken by a single carrying rail with lower guide rails supported on braced trestles three feet high, deeply sunk and eliminating the cost of a traditional roadbed, ballast, and sleepers. Drawbridges lifted the rail clear at road crossings and a pivoted rail was used at turntables and switchpoints. Locomotive, wagons, and carriages straddled the center rail. The locomotives had twin boilers, cabs, and coal bunkers. Two cylinders powered three coupled axles with 24-inch-diameter wheels between the boilers. The tender, on two coupled wheels, had two cylinders to provide additional power. Passengers sat back to back, and the train guard had to ensure that the train had a balanced load before signalling the "right-away." To take a cow to market a farmer had also to bring along two calves, which balanced the cow on the outward journey and each other on the return. With an operating average speed of 18 mph the railway ran successfully for 36 years until 1924.

Smoke and steam generated by the locomotives of the Metropolitan Railway, the first urban underground line, were partially dispersed by occasional open cuttings in this very shallow subterranean system. Clearly, deep-level tube tunnels could not be considered for transport using steam traction. In 1864 a pneumatic railway was demonstrated at the Crystal Palace, the carriage being puffed through the tube like a pellet through a gun barrel. This entirely practical exploit in the tradition of Jules Verne (who was in top form at that time) was not taken up. The deep-level tube railway had to await reliable electric traction. The first underground urban electric transit system was the City & South London Railway, running three miles between King William Street in the City and Stockwell on the other side of the Thames, and opened on December 18, 1890. The trains were hauled by four-wheel locomotives operating on a 600-volt DC power supply generated in the company's own power station at Stockwell. Later, motive power was built into the cars. Because there was no scenery, the original passenger cars were constructed without proper windows (and were nicknamed "padded cells"), the names of the stations being shouted down the cars by the conductor.

A true monorail, highly successful since its establishment in 1901, is the *Schwebebahn* (swinging railway) at Wuppertal in Germany, built by Engen Langen to serve the two industrial communities of Barmen and Elberfeld in the Ruhr. It is essentially a passenger line, on the single rail suspended system, with a maximum operating speed of 31 mph. The line is 9.3 miles long, of which 6.25 miles is suspended directly over the Wupper River, with the remaining distance running above the streets. The track over the river, which includes some stations, is supported by latticed box girders with every sixth girder in the form of a double "A" frame for longitudinal stresses, bridged by horizontal steel plate girders supporting the rail, which is 39 feet above the water. The electric railcars are suspended by a hook-like construction with a pendulum effect, supported by two two-wheeled bogies, with the 59 hp, 600 volt DC motors over the wheels. The 1901 cars, operating as articulated pairs with a passenger capacity of 50 for each unit, remained in use until 1951 and a specimen train is preserved. The line has a remarkable safety record, and its mode of construction made it comparatively easy to repair after heavy wartime bombing attacks.

Escalator, a contrived word invented by Otis Elevator Company to describe the "moving stairway" they had developed with Charles D. Seeberger, was rejected as a noun by the United States Patent Office in the official specification of 1892. It was accordingly registered as an official trademark in 1900. Half a century later, the patent office reversed its thinking and ruled that, because the word *escalator* was in common usage, the noun created by Otis could henceforth be placed in the public domain. Matching the prestige they gained from installing the awkward elevators in the Eiffel Tower 11 years previously, Otis installed the first step-type escalator for public use at the Paris Exposition of 1900. In the following year it was refitted at Gimbel's department store in Philadelphia.

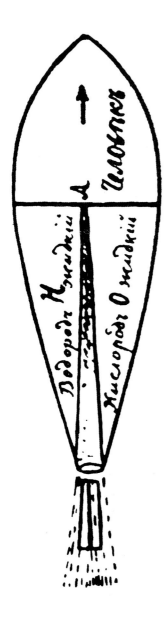

In the same year that the Wright brothers made their breakthrough into powered flight, the practical problems of interplanetary travel by rocket flight were being tackled in the Moscow magazine *Science Survey*. The author, Konstantin Eduardovich Tsiolkovsky, was a deaf schoolteacher, geographically and mentally isolated in the Russian countryside. He was, in fact, intellectually smothered, for his 1903 article was first submitted to the scientific review in 1898. It postulated a rocket reaction motor, liquid propellants to provide exhaust of sufficiently high velocity, and oxygen supplies for spaceship crews which would be renewed on long voyages by plant life. Now universally regarded as the father of astronautics, Tsiolkovsky remained in obscurity until beyond his 60th birthday, but after the Russian Revolution he was made much of by the maturing schools of scientists. Eventually he was held in such high regard that the Russians determined to celebrate the centenary of his birth by launching the world's first man-made satellite, Sputnik 1, on September 17, 1957. Owing to technical hitches, they were a month late. His early design for a spacecraft, with the crew in the nose capsule and liquid propellant feeding into the combustion chamber, seems so right that it is difficult to appreciate that it was conceived 60 years before its time.

The Wright brothers, Wilbur (1867–1912) and Orville (1871–1948), in their Flyer 1, were the first men to take off in an airplane under its own power, to make powered, sustained, and controlled flights, and to land on ground as high as that from which the aircraft had taken off. These flights were made from a plain in the Kill Devil Hills, North Carolina, on the morning of December 17, 1903. At 10.35 a.m. Orville Wright flew 120 feet in 12 seconds. Next, Wilbur flew 175 feet. Then Orville flew 200 feet. Finally, at noon, Wilbur flew for 59 seconds, covering 852 feet, but achieving over half a mile in air distance against a wind of 20–27 miles per hour, thus reaching a speed of 30 miles per hour. Their machine was a biplane with a wing span of 40 feet four inches, wing area 510 square feet, the wings set with slight anhedral and a camber of one in 20. The length of the machine was 21 feet one inch and its gross weight about 750 pounds. It was powered by a 12 horsepower water-cooled gasoline engine designed and built by the Wrights with their mechanic C.Taylor; it weighed 200 pounds and drove two pusher propellers, also designed by the Wrights, with a diameter of eight feet six inches. The Wrights achieved flight in four years, beginning with a biplane kite in 1899 and using their system of flight control by wing warping, and developed three biplane gliders in which they flew up to $622\frac{1}{2}$ feet before building Flyer 1.

In 1877 the 29-year-old Italian engineer Enrico Forlanini had built a successful steam-driven model helicopter on what were then more conventional mechanical principles. Two large contrarotating rotors were worked by a steam engine. But Forlanini had cunningly saved much weight by having a separate boiler which was firmly earthbound. He heated the boiler until he had the required pressure, then introduced the steam to his engine and let go. As a result the three and a half kilogram helicopter soared to over 10 meters and stayed in flight for 20 seconds. Forlanini's most practical work in aeronautics was as an airship designer. His *Leonardo da Vinci*, built in 1909 with steam propulsion, was afterwards converted for gasoline engines. He designed many airships for the Regia Aeronautica during World War I, but he was also the originator of the hydrofoil boat, which he developed between 1898 and 1905. By 1911 he had notched a speed of 77 kilometers per hour at Porto d'Anzio.

Paul Cornu, a French engineer from Lisieux in Normandy, made the first free helicopter flight at Lisieux on November 13, 1907, rising about 30 centimeters, or one foot. His machine had twin rotors, each six meters in diameter, mounted fore and aft on outriggers. The air activated by the rotors played against variable planes which were adjusted for lifting and for directional propulsion. The engine was a 24-horsepower water-cooled Antoinette perfected the previous year by Leon Levavasseur. Forty-five days before Cornu's ascent a helicopter designed by Louis Breguet and Charles Richet had lifted off at Douai, but had been stabilized by four men holding poles under the rotors.

Glenn Hammond Curtiss was, at 29, already the fastest man in the world – having ridden a motorcycle over a measured mile at 136.3 miles per hour – when he joined the Aerial Experiment Association at Hammondsport, New York, in 1907. He installed his 30–40 horsepower V8 air-cooled engine in some of the Association's machines. At the famous Reims Air Week in August, 1909, Curtiss won the Gordon Bennett Trophy in his Golden Flyer pusher biplane at an average air speed of 47.09 miles per hour. Almost immediately he began to concentrate on naval flying and on seaplanes. On November 14, 1910, Eugene Ely, flying a Curtiss biplane of the Golden Flyer type, took off from a small sloping platform (28 feet by 83 feet) which had been built onto the United States cruiser *Birmingham*, and landed on shore at Hampton Roads, Virginia. On January 18, 1911, Ely took off from an airfield near San Francisco and landed on board the United States cruiser *Pennsylvania*, using a hook to grapple stretched cables as his arrester gear. Later in 1911 Curtiss fitted his biplane with floats and inaugurated the system of hoisting seaplanes aboard naval vessels and launching them into the sea for take-off. Curtiss seaplanes were the only American-built aircraft to go into action in World War I.

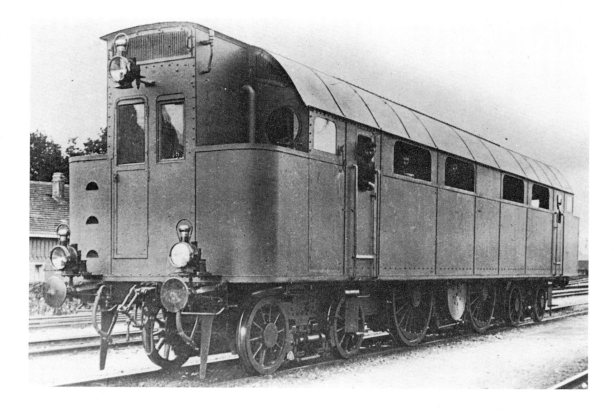

Rudolf Diesel's installation of a diesel engine into a locomotive of the Prussian state railroad system in the winter of 1912 was an historic marker towards the eventual supremacy of diesel power in rail transport. Unfortunately, Diesel did not see it in this light, and shortly afterward threw himself into the English Channel from a packetboat. He had put 20 years of theoretical work into his engine, in the hope of producing a prime mover for economical use in small workshops with better thermal efficiency than the steam engine. It was to be an internal combustion engine burning cheap fuel ignited only by the high temperature of greatly compressed air fed with the fuel into the cylinder. He had to abandon much of his theory, but he did produce an engine of high thermal efficiency. Its "diesel-electric" adaption to drive generators supplying current to tractor motors on the axles of railway trains was revolutionary. Union Pacific used diesel power for a prestige passenger train in 1934, but the outstanding modern impact has been in the rail freight service.

Mass-produced automobiles were first manufactured in Detroit between 1901 and 1904 when Ransom E. Olds began manufacturing 5,000 light one-cylinder Oldsmobiles a year. High precision of interchangeable parts was most strikingly demonstrated by Henry M. Cleland, who in 1908 had three of his Cadillacs disassembled at Britain's Royal Automobile Club. He scrambled all the parts, and let club officials replace from Cadillac stock 90 parts chosen at random. The cars were reassembled, and promptly completed a 500-mile test run. Between these extremes of the automobile market, Henry Ford worked at his great dream to mass produce *the car for the great multitude* and sell it at $600. In 1913 production of the Ford Model T, the "Tin Lizzie" already five years old, was converted to the method, using the band conveyor moving assembly line. Duration of assembly was dramatically cut, and by 1914 production had almost doubled to 300,000 cars a year. In 1924 two million Model Ts were produced and the price was down to $290. The Germans named the new assembly-line technique *Fordismus.*

In 1914 Winston Churchill, then First Lord of the Admiralty, commissioned the adaptation of an American Holt tractor using the British patent of the caterpillar track into a vehicle which could cross trenches. The War Office was aloof, and the Admiralty pressed for its application as a "land battleship" to be operated by the Royal Naval Air Service, which was then operating effective military backup in Belgium. The machine was originally designed by a RNAS lieutenant, W.G.Wilson, working with the engineering firm Wm. Foster of Lincoln. It was converted from its "Little Willie" water-tank shape of September 1915 to a lozenge silhouette, and was finally ordered by the War Office. General Sir Douglas Haig, against expert advice, committed tanks into action at Fler during the battle of the Somme in September, 1916. The ground had already endured a three-day bombardment from 1,000 guns and most of the tanks were bogged down. Having shed the rear wheels, initially used to aid steering, tanks were used effectively at Cambrai on November 20, 1917. The Germans supremely exploited the tank in *blitzkrieg* armored panzer columns in World War II.

The autogiro (a brand name for the pioneer make of gyroplane) was an aircraft with rotors which were not powered, but which were free and "automatically girated" with horizontal motion. They provided lift but no propulsion – the machine being driven by a conventional airscrew and mainly controlled by conventional aircraft surfaces, including an intact tail unit. It could take off and land on extremely short runways, but could not hover or move sideways or backwards. The machine was developed by Juan de la Cierva, who on January 9, 1923, made his first successful flight at Getafe, near Madrid, in a Cierva C.4 powered by a 110-horsepower Le Rhône rotary engine. From 1925 Cierva researched and manufactured in England. His work greatly advanced research on the helicopter; his introduction of flapping blades was a vital contribution, and he almost equalled helicopter potential in his C.30 model by power-spinning the rotor before take-off to achieve a jump start.

On March 16, 1926, Robert Hutching Goddard, a 43-year-old American physics lecturer, erected a light launching stand in open country at Auburn, Massachusetts, and set up a rocket in it. This slim construction was some 10 feet high, and carried in its nose a motor two feet long. Its combustion chamber was fed from tanks containing liquid oxygen and petrol. Goddard lit a blowlamp fastened to a pole, opened the fuel tanks, raised the pole with the flaming crest like an old lamplighter, and applied it to the igniter. The fuel roared, the rocket blasted off, the propellant was consumed in two and a half seconds, and the rocket fell to the ground after reaching a height of 184 feet at an average speed of 60 miles per hour. This was the first successful liquid-fuel rocket ascent, and the precursor of the flights that would put men on the Moon. Goddard had been working quietly on rockets for 20 years, always with a view to their eventual entry into space, but with the immediate objective of high-altitude meteorological research. He later launched rockets in New Mexico up to a height of one and a half miles, constantly publishing his research, and taking out 214 patents. Although a few side products were taken up for military purposes, his main work was ignored by the United States Government. But in 1945, when Goddard died, Wernher von Braun and others who were being brought over from Germany to start work in what became NASA told the Americans that virtually all the rocketry they knew had come from Goddard's work.

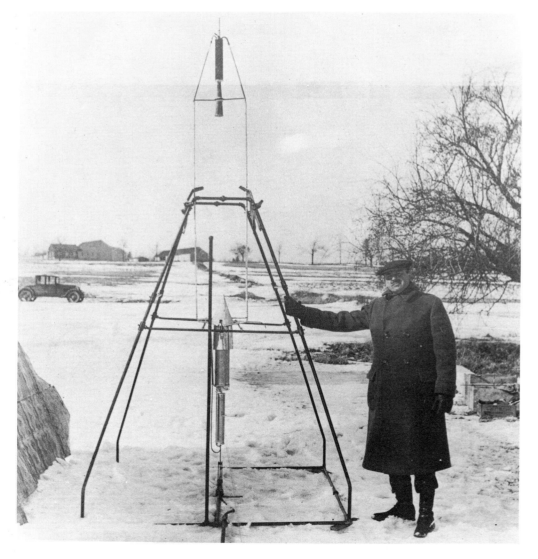

234

On June 11, 1928, Fritz Stamer made the world's first manned flight by rocket propulsion in an Opel-Stamer glider of the *canard* configuration used by Santos-Dumont over 20 years before. He flew for 1,220 meters (about 4,000 feet). The machine was virtually uncontrollable and made no further significant flights. The Opel-Stamer machine was a cul-de-sac enterprise occasioned by the brief publicity venture into rocket propulsion by the rich German automobile manufacturer Fritz von Opel. It was an offbeat development of the interest in rocketry promoted by the German VfR (*Verein fur Raumshiffahrt*, Society for Space Travel) formed in 1927 and disbanded in 1933. The VfR was the most effective national society of space travel and interplanetary enthusiasts between the wars. The man who later emerged as the most famous member of the old VfR was Wernher von Braun.

The U.S. Navy airship *Shenandoah* was the first airship to take its lift from the inert gas helium, which is about twice as heavy as hydrogen and which in 1923, when the aircraft was built, could be obtained in quantity only in North America. Unfortunately, though the fire risk was reduced, the design of the aircraft was still vulnerable. In 1925 the *Shenandoah* broke in three parts over Ohio. This was only one casualty among the many between-wars disasters affecting large rigid airships. Others were the loss of the British *R38* in 1921 and *R101* in 1930; the United States *Akron* in 1933; the French *Dixmude* in 1923; the German *Hindenburg* in 1937. These ruinous losses – never attributable to one reparable design fault – led to the abandoning of large airships and even the withdrawal of the totally reliable *Graf Zeppelin*, the most successful passenger-carrying airship in history. Between 1928 and 1937 she made some 600 commercial voyages, including 144 ocean crossings, was in the air for at least 17,000 flying hours, carried more than 13,000 passengers, and traveled at least one million miles, all without an accident. The *Graf Zeppelin*, the company's 127th airship, was named after the founder Count Ferdinand von Zeppelin; it began its maiden flight, from Friedrichshafen to Lakehurst, New Jersey, on October 11, 1928, with 20 passengers and 40 crew. Later it flew round the world in three weeks, of which 12½ days were flying time. Between 1932 and 1937 the airship fulfilled a regular passenger service across the south Atlantic, flying from Friedrichshafen to Rio de Janeiro. Passenger accommodation was far more spacious than in any modern Jumbo.

The theoretical and practical limits to the use of the optical microscope were set by the wavelength of light. By the time the oscilloscope was developed, it was realized that cathode-ray beams, manipulated by magnetic and electrostatic fields, could be used to resolve much finer detail because their wavelength was so much shorter than that of light. In 1928 Ernst Ruska and Max Knoll, using magnetic fields to "focus" electrons in a cathode-ray beam, produced a crude instrument which gave a magnification of 17. By 1932 they had developed an electron microscope with a magnification of 400. James Hillier's expertise advanced this magnification to 7,000 by 1937; Zworykin's 1939 instrument was to give results 50 times more detailed than any optical microscope. Magnification of up to 2,000,000 has now been achieved. The electron microscope has revolutionized biological research, especially investigation of cell structures, proteins, and viruses, where for the first time scientists have been able to "see" molecules.

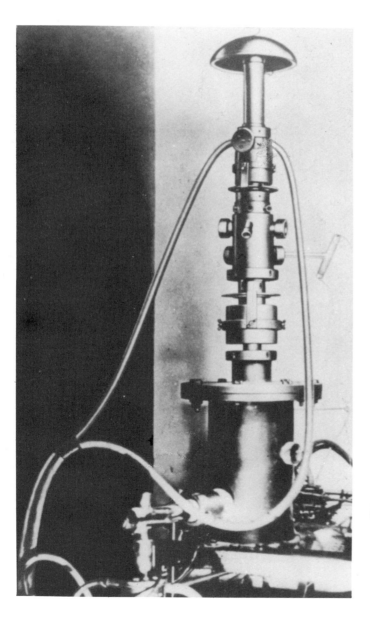

Cathode rays were named in the 1870s, and were studied with increasing authority from that time. William Crookes was the first to define and exploit this radiation from the negative electrode of an evacuated tube subjected to a high-voltage current. Nuclear physics, X-rays, radiocommunication, and television are only a few of the fruits of that harvest. Theoretically, there could be television today without dependence on the cathode-ray tube. John Logie Baird was, well into the 1920s, following the ingenious system patented by Paul Nipkow in 1884, using a mechanical scanner. But the resolution and sensitivity of mechanical optical devices was inadequate. An electron-scanning tube had existed since 1897, the work of Karl Ferdinand Braun, who had modified a cathode-ray tube in an oscillograph. Braun's work was well known to physicists. In St. Petersburg, Professor Boris Rosing had experimented with a cathode-ray oscilloscope as a television receiver, but using a mechanically scanning transmitter. His student, Vladimir Zworykin, who emigrated to the United States in 1919, was determined to make an *electronic* transmitter, or what we now call a television camera. In 1928 the year of the first "all-talking" feature film, Zworykin produced his iconoscope, a camera tube which stored on individual photocells the light that had fallen on them when they were exposed, and which released this store of information when "switched on" by a cathode-ray beam. With speed astonishing for a bureaucracy, the British Broadcasting Corporation introduced public television to the world in 1936, using Zworykin-type apparatus. Zworykin later instituted the important modifications which made the electron microscope a practical instrument.

In 1931 Karl Jansky was investigating for Bell Telephone Laboratories the crackling "static" which interfered with radio-telephony. He timed a periodic interference in short-wave reception at a cycle of 23 hours 56 minutes, the exact frequency of the Earth's daily rotation in relation to the stars, not the Sun. He concluded that a source of radio bombardment was in the constellation of Sagittarius in outer space. He had stumbled on the potential of radio astronomy, the reception and interpretation of radio microwaves – radiated not only from Sagittarius but from the entire galactic plane. Because they pass through dust clouds which optical telescopes cannot penetrate, the microwaves give a new dimension to space investigation. Jansky's discovery was taken up by an amateur astronomer, Grote Reber of Illinois, who set up the first radio telescope in his garden in 1938. In 1951 Manchester University established the first chair of radio astronomy with Bernard Lovell as professor. Lovell began construction of the great Jodrell Bank "dish" radio telescope, which promptly justified itself by its accurate tracking of Sputnik I.

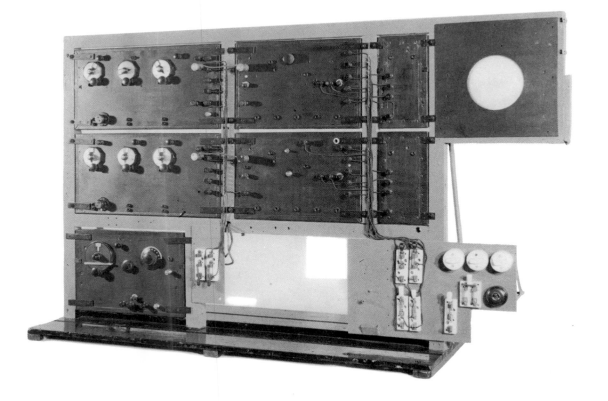

RADAR, standing for Radio Detection and Ranging, was not so named until World War II, but had been feverishly developed by Great Britain as a defense technology through the 1930s. Robert Watson-Watt of St. Andrews, Scotland, had specialized in the echo reflection of radio waves. The invention by Albert Hall in 1921 of the magnetron vacuum tube permitted the generation of radio microwaves with wavelengths of less than 50 centimeters. Watt concentrated on the accurate measurement of the reflection of microwaves generated in this manner. He succeeded in measuring and interpreting the deflection given to these electromagnetic waves by objects in their straight path. Radar observation was not impeded by cloud, fog, or darkness, and Watt was able to detect and plot unseen distant objects; the emphasis at that time was naturally on the identification of enemy aircraft and, later, submarines. By 1935, largely through the energy of Henry Tizard, Great Britain had five operational radar stations, and the total was 20 before the war came. Radar was subsequently developed to construct the proximity fuse, the device in the warhead of a projectile which detonates the charge at an optimum distance from its target.

The pioneer aircraft of all modern airliners, the Boeing 247, made its first test flight from Seattle, Washington, on February 8, 1933, and entered commercial service with United Air Lines in March of the same year. The Boeing 247 revolutionized and standardized all long-distance commercial air transport. It was a low-wing cantilever monoplane with a 74-foot wing span. Originally its two Pratt & Whitney R-1340-S1H1G Wasp nine-cylinder radial engines, gave it a cruising speed of 155 miles per hour with a range of 485 miles. Many refinements, including controllable-pitch propellers, gave a cruising speed of 189 miles per hour at 12,000 feet and a range of 745 miles. The Boeing 247 had a crew of two, accommodation for 10 passengers, and storage for 400 pounds of mail. It was speedily followed into commercial viability by the Douglas Commercial DC range, culminating in the famous DC-3, and the two designs set the pattern for modern all-metal aircraft with retractable undercarriage.

The Sikorsky VS 300 helicopter, "the first fully practical and successful single rotor configuration machine," finally satisfied its designer and went into production in 1942, although it had been held off the ground at least long enough for this photograph on September 14, 1939. Igor Ivan Sikorsky was born in Russia in 1889, the son of a distinguished professor of medicine. His mother, also a doctor, fascinated her son with stories of Leonardo da Vinci and from boyhood he was intrigued with helicopters. In 1910 he built one in Kiev, using a 25-horsepower Anzani engine. It did not get off the ground. Sikorsky turned to fixed-wing aircraft and his huge "Bolshoi" biplane, powered by four 100 horsepower Argus four-cylinder engines and carrying eight passengers in a "greenhouse" cabin glazed with unbreakable glass, successfully took the air in 1913. After the Russian Revolution Sikorsky emigrated penniless to the United States. He began an aircraft company from scratch and developed it to make planes and flying boats. But he constantly reverted to his conception of the helicopter, relying on the anti-torque rotor at the tail, and striving for full cyclic pitch control with one main rotor. (With cyclic pitch control the angle of attack of the rotor blades is changed to produce a horizontal thrust component, allowing the helicopter horizontal flight in any direction.) As a consequence of his determination, Sikorsky's name is now almost synonymous with helicopter.

Gas turbines were always a theoretical alternative to piston engines in aircraft, capable of harnessing the energy released by fuel combustion but wasted in the exhaust, and of overcoming the increasing inefficiency of piston engines at high speeds and high altitudes. The turbo prop system, an early proposal, would have entailed having the turbine drive an air screw. Frank Whittle envisaged the turbojet, using the flow of hot gases from an engine to collect and compress additional air, fed with injected fuel into a combustion chamber to produce hot gas escaping with increased energy. Whittle first advanced this idea when he was a 21-year-old RAF cadet in 1928. It was rejected, and he patented it privately. By April 12, 1937, he had built and successfully tested his first turbojet engine. In Germany Pabst von Ohain, sponsored by Ernst Heinkel, tested his first jet engine in September, 1937. Though both were working without government backing, Heinkel's progress was faster than Whittle's, and on August 27, 1939, Erich Warsitz flew the first jet aircraft, the Heinkel He 178, with Ohain's He S3B engine. The British Gloster E28/29, with a Whittle W1 engine, flew first on May 15, 1941.

Ten weeks before flying the He 178, Warsitz successfully piloted the He 176 rocket plane. Rocket propulsion was as old as gunpowder, and military minds bogged down in the gunpowder age. International space exploration enthusiasts – whom the military regarded as cranks – switched practical thinking to liquid fuel propellant. The work of Tsiolkovsky in Imperial Russia, Goddard in the United States, and Oberth in Rumania and Germany was encapsulated and developed by Wernher von Braun, an original member of the German Society for Space Travel who was side-tracked into ballistic missile research when the German army took over the rocket program in 1932. Braun's V2 was successfully test-fired on October 3, 1942. The first of more than 4,000 V2 rocket missiles was fired on London at 6.36 pm on September 8, 1944. The V2 was a 46-foot-long, four-ton deadweight shell powered by five tons of liquid oxygen and four tons of a three-to-one alcohol-water mixture, carrying 1,650 lbs of Amatol explosive in its warhead, with a range of 200–220 miles and a maximum speed of 3,600 miles an hour. After the defeat of Germany it was developed in the United States by von Braun and in the Soviet Union by his colleagues. Russia, with inferiority in nuclear technology, concentrated on the rocket, and by 1957 had launched an intercontinental ballistic missile with a range of 5,500 miles and the satellite Sputnik I, which the world saw as it orbited the Earth.

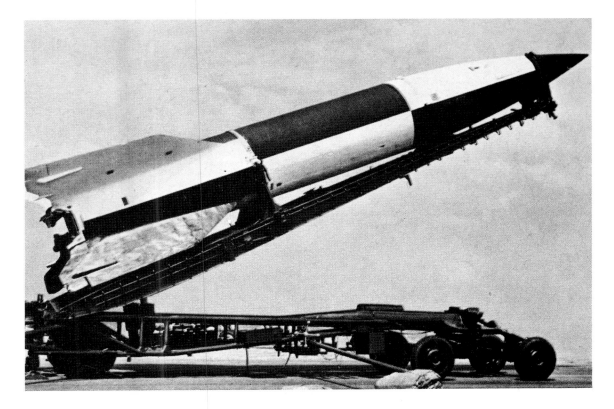

Nuclear energy is the capacity for work which is locked within the nucleus of the atom. During the half-century following Sir William Crookes' work on cathode rays in 1875 the atom – nominally indivisible – was shown to be divisible and was seen as a solar system of negatively charged particles (electrons) orbiting around the nucleus of positively charged particles (protons) and uncharged particles (neutrons). To fragment the nucleus (popularly known as splitting the atom) would release energy, but when this was done by Lord Rutherford in 1919 and by Sir John Cockcroft in 1932 more energy had to be applied to the nucleus than was released by it. These scientists had bombarded nuclei with protons but were restricted to a hit-or-miss technique of the order of one hit to a million shots. In 1935 Leo Szilard, an émigré anti-Nazi physicist working in Britain, theoretically projected nuclear chain reaction by the cumulative production of neutrons which would split further nuclei. Meanwhile in Italy Enrico Fermi, a quiet anti-Fascist professor of physics in Rome, developed the bombardment of nuclei of uranium, then the heaviest element known, with neutrons which he slowed down so that they hit their target more frequently. When Szilard and Fermi arrived as political refugees in the United States at the end of the 1930s, Szilard set Fermi to the problem of converting a ''one-off'' release of energy into a cumulative chain reaction, using uranium research developed in England by Otto Frisch and Rudolf Peierls. Fermi set up in Chicago the world's first ''atomic pile,'' a controlled continuous source of energy rather than a hit-or-miss reaction. Between 3 and 4 pm on December 2, 1942, the pile was producing cumulative energy by chain reaction and a coded telegram was passed: *The Italian navigator has entered a new world*.

As Fermi had forecast, his nuclear reactor manufactured a further fissile element as a by-product from uranium: plutonium. During the next 30 months, when every effort was channeled to apply nuclear energy for destruction rather than conservation, the United States spent billions of dollars on two tasks: the manufacture of sufficient plutonium and enriched (''bomb quality'') uranium for three bombs and their preliminary tests, and a solution to the problem of getting an atomic bomb to explode at the right time. The first nuclear bomb, using uranium, was tested from the top of a steel tower at the Trinity site at Alamogordo, New Mexico, on July 16, 1945, and was dropped on Hiroshima 20 days later on August 6, 1945. The plutonium bomb, tested in the Nevada desert soon afterwards, was dropped on Nagasaki on August 9, 1945.

The mechanical origin of the computer is the odometer described in the book attributed to the Greek engineer Hero of Alexandria in the 1st century AD. The odometer contained a pin in the hub of a wheel on a moving carriage which worked a train of wheel-and-worm gears. These, in turn, operated dials which showed in various magnitudes (like the dials on domestic power meters) the number of revolutions of a carriage wheel and the total distance covered. This principle was adapted by Blaise Pascal for his adding and subtracting machine built in 1642, which was in turn improved by Leibniz between 1671 and 1694, to tackle multiplication and division. These machines were primitive *digital* calculators dealing with integral numbers like separate balls on an abacus. The manual slide rule already existed, using measurements proportionate to the logarithms of numbers as a very crude analog "computer." Between 1822 and 1871 Charles Babbage, having made one calculating machine while laboring to correct errors in logarithm tables, was working on his ambitious Analytical Engine. It was intended to store and retrieve information submitted by punched cards, and to print out its answers. Technical competence was inadequate at that time, and the first mechanized analog

calculator was the Differential Analyzer built by Vannevar Bush at the Massachusetts Institute of Technology in 1925. At first its function was limited to the solution of differential equations. While this was being elaborated, Howard Aiken at Harvard adapted Babbage's punched-card information system, (which had been developed by Hermann Hollerith). He applied it to various mechanical sequences which worked out arithmetical presentations. With a number of experts from the International Business Machines Corporation (the successors to Hollerith), he constructed the IBM Automatic Sequence Controlled Calculator, better known as the Harvard Mark I computer, which was put into operation in August, 1944. It was a mechanical mathematical machine using electric power, and it stood eight feet high and 51 feet long. The first electronic computer, the United States Army's ENIAC (Electronic Numerical Integrator and Computer) was even more unwieldy. It was not mechanical in operation, but worked by pulsed circuitry, as first developed in radar. ENIAC contained 18,000 radio tubes, or valves, and demanded a power input of 130 kilowatts. Computers became more manageable in scale after the introduction of transistors in 1948.

William Shockley and his research colleagues at Bell Laboratories, John Bardeen and Walter Brattain, worked on the capacity of certain crystals to act as rectifiers (passing alternating current in surges in only one direction), and found a method of combining solid-state rectifiers so that the device also amplified current. This development, called a transistor because it transferred current across a resistor, could do all that a radio tube did; it could be made to do it better and without preliminary heating-up, and was infinitely smaller and handier. The first transistor ever made was a point contact device in which the emitter and the collector were attached to a block of germanium crystal mounted on an electrode. The bias on the emitter produced amplification by controlling the current between the collector and the base mounting. Further work on the transistor, including the "sandwich" junction, brought Shockley, Bardeen, and Brattain a Nobel Prize. Without the transistor, the gargantuan laboratory computer would have become virtually calcified by the increasing demands on it. Moreover, space research and space travel, as well as spatial aggression, would have been impractical without the portable miniaturization afforded by transistors.

Enrico Fermi's "atomic pile," the first nuclear reactor, had given an output of about half a watt when it was activated in December 1942. Since Fermi was then involved in developing the uncontrolled eruption of energy in an atomic bomb, the output was irrelevant though still significant. After the war some scientists were anxious to demonstrate that nuclear fission could be used to generate power for peaceful purposes – such as cooking, heating, and lighting. The first full-scale nuclear power station, laid down at Calder Hall in Cumbria, England, from August, 1953, and in operation from October, 1956, was a slightly embarrassing compromise. It was designed to produce not only power but plutonium, which at that date was a nuclear weapon; by having two functions, it did neither of them with optimum efficiency. The first generation of British nuclear power stations, now obsolete in terms of capital cost against return, produced about one-tenth of the national requirement of power.

At Oak Ridge, in the United States, Captain (later Admiral) Hyman G. Rickover masterminded the decision to aim for nuclear-powered submarines using available fissionable materials. It was a delicate decision, at a time when no one knew how far the Korean war would escalate, and necessitated the use of proven technology rather than further extended research. Under these peculiarly makeshift conditions which also governed the decision on Calder Hall, Rickover designed the first nuclear submarine *Nautilus*, built at Groton, Conn., and launched on the River Thames there on January 21, 1954. Almost a year later, on January 17, 1955, the 324-foot-long, 3,747-ton (submerged displacement) craft signalled "under way on nuclear power" during its sea trial. Two years later, after an aggregate of some 200 miles a day, it went in for first refueling. *Nautilus*, the first of many nuclear-powered submarines, was later equipped with the Polaris solid-fuel rocket, carrying an intermediate range ballistic missile which could be launched underwater. With a theoretical potential to "disappear" for two years carrying this nuclear weapon, it and its consorts forced a new dimension into the global weapons array.

Enrico Forlanini, the airship designer who included helicopters and hydrofoils among his side interests, demonstrated his 77 km/h hydrofoil of 1911 to Alexander Graham Bell, who improved it and took the water-speed record to 71 mph in 1918. Subsequent developments were made by German engineers and by a British organ builder, Christopher Hook. In 1952 in Messina, Sicily, Carlo Rodriquez exploited the designs of Baron Schertel von Sachsemberg and the skills of a number of German technicians who had been prisoners of war in Russia, to build and operate the first commercial hydrofoil service. He sold the Supramar concept in many parts of the world. Hydrofoil construction overcomes drag by lifting the hull clear of the water, still supporting it on skis or foils. In commercial practice it is the fastest passenger craft on water.

On July 25, 1959, the 50th anniversary of Louis Blériot's flight across the English Channel, Christopher Cockerell flew across the same stretch at much the same speed in SRN-1, a bulky passenger-carrying hovercraft traveling only a few inches above the surface of the water. Cockerell, a boatbuilder wrestling with problems of drag, developed an experiment whereby he hit upon the use of a cushion of air between hull and water. Working at home, he had blown air from a vacuum cleaner motor downwards onto kitchen scales through discarded tins of different diameters. He found that, contrary to the laws of physics as he understood them, air blown through a narrow jet exerted disproportionately increased thrust compared with the same mass of air blown through a wider coffee can. Utilizing this unexpected lift, he built a model hovercraft with a "skirt" or curtain to contain the air giving the thrust. He innocently took the idea to the Admiralty, which shoved it on the secret list and forbade Cockerell from exploiting it. Eventually the ban was lifted and Cockerell was knighted. The air cushion principle was later tried for fast rail transport, but was abandoned on cost grounds.

243

The idea of harnessing the power of the tide, involving the regular disturbance of all fluids on Earth by the gravitational forces of the Sun and the Moon, is not new. Tide mills were known since the Norman Conquest and used while the miller found it economic to work unsocial hours. The first significant tidal power station for the generation of electricity was set up between 1961 and 1966 across the Rance estuary in the Golfe de St. Malo off Brittany in France. In a combination of engineering daring and elegance, supported by brilliant mathematics, this construction uses the tide to fill a reservoir with additional hydroelectric function and has 24 turbo-alternator sets in midstream which can operate whatever the direction or intensity of the tide and can be worked as turbines, pumps, or open sluices. In this way, the half-mile Rance barrage flexibly and economically links the cyclic demand for peak power with the remorseless rival cycle of the tide which pays no heed to human dinner time. The annual output is over 500 million kilowatt-hours.

The amount of energy shed by the Sun on the Earth during two weeks is equivalent to the world's entire initial stock – now fast falling – of coal, oil, and gas. Every passing day makes the collection and utilization of inexhaustible solar energy more economical, and its exploitation carries no danger of the upset of terrestrial energy balance which haunts all schemes to exploit nuclear reaction.
The French solar furnace at Odeillo in the Pyrenees masses 11,000 adjustable mirrors on a south-facing mountain to reflect the Sun's light onto 9,000 mirrors across the valley, set up in a parabolic curve to re-reflect the Sun's light focused onto a furnace room which, almost as soon as the Sun shines, reaches a temperature of 3,300°C.

THE SPACE AGE

The manned spaceflight missions from 1969–1972 capped a technological enterprise that grew in a few decades from a technical hobby into the complexities of mobile television studios operated by men on another world. A successful integration of many other new technologies (computers, micro-electronics, radio communications, ceramics) led to the most fantastic voyage in the history of exploration: manned landings on the Moon, machines landed safely on nearby planets, and probes sent to the outposts of the solar system and the beckoning cosmos beyond.

To get into space it is necessary to escape the gravitational field of the Earth. A material object is freed from Earth's grasp if it attains a velocity of 25,000 miles per hour.

However, relative to everyday technology, Earth's escape velocity is large: 40 times that of a jet plane. An important consequence of this high velocity is that the propellant requirements of a space rocket are huge, constituting up to 95 percent of the rocket and space probe prior to launching. The technological achievement entails burning the fuel smoothly in a relatively short time to achieve the desired acceleration. In the Saturn 5 boosters used in the u.s. manned spaceflight program, maximum thrust of more than seven million pounds occurs 90 seconds after launch. Within 10 minutes the rocket is parked in orbit at a height of 100 miles. Although practical propulsion systems based on nuclear fuels eventually may be developed, rocket technology until now has been limited to systems in which the initial fuel mass far exceeds the payload mass.

Overleaf: illustration from Jules Verne's *Journey to The Moon,* 1865

The Chinese are credited with the discovery of gunpowder and the application of it to pyrotechnic devices about 1,000 years ago. For the next nine centuries most advances were made by theorists and science fiction writers. Sir Isaac Newton, in developing the theory of gravitation, correctly concluded that a projectile launched with sufficient speed would orbit the Earth continuously. Daniel Defoe speculated in *The Consolidator* (1705) that the Chinese had received the secrets of space travel from extraterrestrial visitors. Jules Verne was the first to conceive the technological problems correctly. In *From the Earth to the Moon* (1865) he describes a space capsule with hydraulic shock absorbers, launched from a cannon by 400,000 tons of explosives. Interestingly, Verne starts the voyage in Florida and returns his three heroes to the Pacific Ocean, exactly as Apollo 11 was to do in 1969.

Attempts to build rockets for research purposes date from the early 20th century – the work of Tsiolkovsky in the Soviet Union and Robert Goddard in the U.S. being particularly significant. The space era truly commenced in the 1930s, when fund-starved groups in Germany and the U.S.S.R. conducted fundamental research on rocket technology. During the 1939–45 war the German group, working at Peenemunde under Walter Dornberger and Wernher von Braun, developed launch systems for the V1 and V2 weapons. Near the end of the war, von Braun and some of their associates surrendered by choice to the Americans, but others in the Peenemunde team joined the Soviets to develop the V2 rocket for the Red Army. German scientists made fundamental contributions to the U.S.S.R./U.S. space race of the 1960s.

International Geophysical Year, 1957, marked the opening of the space age. Soviet and American scientists proposed that an artificial satellite would be of great value for remote surveillance of the terrestrial environment. The successful launch of Sputnik 1, a 184-pound craft, on October 4, 1957, was a shattering blow to American technological prestige, and was the first in a long line of Soviet firsts in space. A two-stage rocket fired in Central Asia had put a fully operational satellite into orbit. Members of Congress dismissed Sputnik 1 as a hoax, while England's Astronomer Royal solemnly declared *space travel is sheer bunk*.

Undeterred by this polemic, the Soviets blasted the first living creature, a dog, into orbit four weeks later. A TV audience of millions saw a U.S. Navy attempt at a satellite launch end in a ground-level fireball. With von Braun's help, the U.S. Army sent Explorer 1, a 31-pound satellite, aloft on a Jupiter-C rocket in 1958. The U.S. Navy got a three-pound satellite up on March 17, 1958. But within two more months the Soviets had Sputnik 3 in orbit, a remarkable achievement, since the craft included one ton of scientific instruments and power sources.

Lunar research in the period 1966–68 was dominated by the use of soft landers and orbiters. Five Soviet moonshots (Lunas 4–8) smashed into the Moon, whereas the Americans gently landed Surveyor 1 in July, 1966. This required the development of rocket systems that could control the speed of final approach. The U.S. spacecraft also employed solar cells to generate electrical power, enabling them to function for many months, whereas Soviet landers had to manage with chemical batteries that ran down after three to five days. By using solar cells rather than batteries, the Surveyors obtained vastly superior scientific data. Surveyor 4 (July, 1967) furnished 19,054 television frames of the lunar surface, stars, the solar corona, and earthshine, providing astronomers with the first measurements of coronal brightness out to 30 solar radii.

The U.S. pressed ahead with the manned program at astonishing speed. The Apollo program, judged in terms of design, execution and political achievement, is one of the chief technological glories of the big-budget science and engineering projects of the 1960s. Manned spaceflight is far more complex, technologically and socially, than unmanned flight. The provision of life support systems and the unacceptability of fatal accidents in space greatly added to the payload and fuel requirements. The Apollo plan therefore had to await the completion of the Saturn C-5 booster, constructed in Alabama under the direction of NASA. This behemoth of rockets contributed in large measure to the superior achievements of the American space program from the mid-1960s onwards. The height of the rocket exceeds 300 feet and the weight at launch is 3,000 tons, about the same as a naval light cruiser. To accelerate this hardware up to a velocity of 25,000 miles per hour demands a thrust of more than seven million pounds on lift-off, equivalent to 150 million horsepower. In the first stage of the flight, 2,250 tons of fuel are burned in under three minutes. This system is able to send a payload of 45 tons on its three-day journey to the Moon.

In July, 1968, the first manned flight around the Moon took place, followed by the historic landing on July 20, 1969. Apollo 15 (July 26, 1971) greatly extended the scope of lunar exploration by transporting to the Moon the first lunar Rover – the most expensive car ever constructed.

In the field of planetary exploration a manned trip would have to last about one year, owing to the distances involved. This is prohibitively expensive at present. The major powers have therefore concentrated on photographic and electronic surveillance and the development of landers.

The late 1970s have been a time of consolidation. No major new devices have been invented, but the techniques used for the inner planets have been adapted for the long-haul routes to Jupiter and Saturn. There are plans to put landers on the larger moons of these giant planets by the 1990s.

For Earth orbit, the technology is now so well-developed that most expenditure is on applications rather than research and development. Among the conspicuous achievements are the use of satellites for weather prediction, remote sensing of the environment, mapping of resources for the distribution of energy, military and defense surveillance, and communications. The high quality of transcontinental TV transmissions and the introduction of automatic international dialing for telephone subscribers has resulted from the use of relay satellites in place of oceanic cables.

The first great era of space exploration, with its unexpected discoveries by pioneering technology, is already ended, and the era of routine applications has commenced.

As part of the International Geophysical Year, stretching from July, 1957, to December, 1958, both the United States and the Soviet Union announced an intention to put observatory satellites into orbit around Earth. The US administration had declared that the Defense Department would not participate in the project, although the three-stage Vanguard vehicle for the 21-pound satellite did, in fact, depend for its first stage on the Navy's Viking rocket which had lifted research instruments to a height of 158 miles in 1954. For the launch of their own satellite, however, the Soviets relied on a war rocket based on Russian and German development of the V2. The intercontinental multistage ballistic missile which the Soviet Union announced in August, 1957, was greeted with disbelief by Western military experts. Two months later, on October 4, 1957, Soviet engineers put into orbit from the adapted missile Sputnik 1. The instrument-packed aluminium sphere was nearly nine times heavier than the projected Vanguard satellite, and the West could hear it as it bleeped its way around Earth every 96 minutes. Seventeen days later the Russians put up Sputnik 2, a vehicle six times heavier than its predecessor and carrying the dog Laika, whose telemetered heartbeats indicated that weightlessness still permitted the bodily functions. In December, 1957, the United States' reply, the lightweight Vanguard, was launched to a height of one inch before exploding into flames.

On February 1st, 1958, Wernher von Braun, working with the United States Army, used his Jupiter-C rocket to put into orbit Explorer 1, which with its altitude ranging from 224 to 1584 miles made the first salient discovery of the space age: the existence, 600 miles up, of the lower Van Allen Belt of subatomic particles from the Sun trapped by the Earth's magnetic field. Vanguard 1 finally got into orbit 45 days later, and the subsequent launchings of Explorers and Pioneers established the American space program, furthered by the inauguration on October 1, 1958, of NASA, the civilian agency for National Aeronautics and Space Administration. By 1960 the 100-foot diameter aluminum-coated balloon Echo 1, in orbit at 1000 miles altitude, was reflecting the signals which made it the world's first communications satellite, relaying transocean television and radio messages.

On April 12, 1961, 27-year-old Yuri Alekseyevich Gagarin, piloting Vostok 1, became the first man to be set in orbit around the Earth and to return alive. He orbited the world once, taking 108 minutes, reaching a speed of 17,400 miles an hour, at altitudes between 112 and 203 miles. Vostok 1 weighed 10,419 pounds. Its length was 124 feet. Its first-stage propulsion was sensational at the time. It was blasted off by 32 rocket chambers firing simultaneously. Around four central primary nozzles and four verniers were four boosters, each with four primary nozzles and two verniers, giving a total thrust of 1,323,000 pounds. To bring the craft out of orbit a retrorocket was fired to reduce speed, and the equipment module was almost immediately jettisoned. Gagarin descended and landed in the re-entry capsule – though his five successors in Vostok spacecraft landed by parachute. Gagarin died in a "conventional" airplane crash on March 27, 1968.

Gagarin had been followed into space within weeks by Alan Shepherd and Virgil Grissom, making short suborbital flights. In 1963 Valentina Tereshkova made 48 orbits within three days. Six years and 21 manned flights later, on July 16, 1969, Michael Collins, Neil Armstrong, and Edward Aldrin were launched by the Saturn 5 rocket in the Apollo 11 space vehicle. Sixty hours later they were orbiting the Moon. At 20:18 hours GMT on July 20 Armstrong and Aldrin in their detached lunar module Eagle, after a controlled descent to the Sea of Tranquillity, reported to Houston: "The Eagle has landed." At 2:56 GMT on July 21, 1969, Armstrong stood on the surface of the Moon and then photographed Aldrin, the second man to land on the Moon, as he made his descent. "We came in peace for all mankind," says a plaque they left to record the event. After a two-hour stint, collecting specimens and setting up research equipment, they slept in their lunar module, lifted off to rejoin the command module Columbia, and splashed down in the Pacific three days later.

In 1789 William Herschel had built the first of the giant telescopes with a mirror 49 inches in diameter. In 1845 Lord Rosse finished his operationally dangerous 72-inch Leviathan. It was to be the world's biggest telescope until 1917 when the 100-inch instrument swung into action on Mount Wilson, California. Today, the value of astronomical observations from Mount Wilson and Mount Palomar has diminished: light that has traveled from almost the beginning of time cannot compete in intensity with the diffused glow of street lights and advertising signs in the urban sprawl of Southern California. The center of astronomy in North America is now Kitt Peak Mountain, 40 miles west of Tucson, Arizona, where dozens of telescopes now sprout from a mountain from which Indians once worshipped the Sun. The 1970 Mayall telescope, with a main mirror of 158 inches, is one of the largest in the world, with a light-gathering power of 200,000 pairs of human eyes. Its array of electronic detectors stores and analyzes the images of faint stars and remote galaxies.

The first truly mobile piece of equipment to be operated remotely on the Moon was the Lunakhod, landed by the Russian probe Luna 17 in November, 1970. The following July, the Americans arrived in style by taking the Lunar Rover with them in the Apollo 15 mission. The Russians responded with Lunakhod 2, weighing almost one tonne at 850 kg, which was landed on January 16, 1973. This seemingly ungainly vehicle trundled around for six months, whereas manned spaceflight missions to the Moon were limited to, at the most, tens of hours. During its period of activity, the Lunakhod 2 equipment withstood temperatures ranging from − 160C in the lunar night to 100C by day. In all, 86 panoramic photographs were obtained, as well as 80,000 TV frames, and the total distance explored was about 28 miles. A later craft in the same series, Luna 24, landed on the Moon, drilled six feet below the surface, took off, and returned safely to Earth with its lunar samples on August 22, 1976.

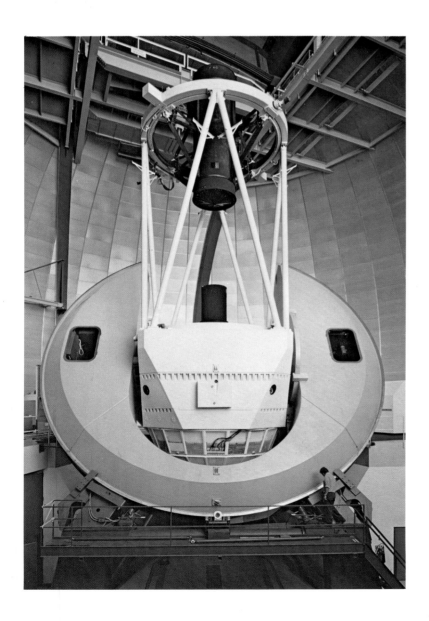

The first Lunar Rover, brought by Apollo 15 in 1971, extended exploration possibilities on the Moon tenfold; in the previous two years the astronauts on the Apollo 12 and 14 missions walked a total distance of three miles. Altogether three Lunar Rovers were taken to the Moon, where they still remain, after traveling a total distance of 50 miles during some 60 hours of activity. The use of rovers vastly increased the scientific value of the projects, for they enabled a much greater variety of rock and soil samples to be gathered and taken back to the lunar excursion module. This capacious vehicle, seen from the command module, held two astronauts, their equipment and supplies for two days on the surface, as well as the Lunar Rover and enough fuel to leave the Moon and dock with the service module and with the command module, in which the whole crew returned to the Earth's orbit. While it was on the surface of the Moon, the lunar excursion module beamed back from TV antennae a steady stream of data and pictures to the base on Earth.

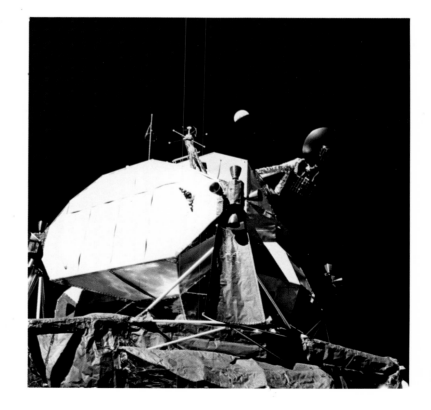

The service and command modules of Apollo 17, orbiting around the Moon in December, 1972, were photographed from the lunar excursion module after separation and before descent to the surface. Throughout the lunar excursion one man remained in the command module to keep contact with the two astronauts in the lander. The exhaust nozzles of the small rocket motors, used for fine control of the attitude of the service module, show clearly, as does the open scientific bay. Cameras in the bay took several thousand photographs, and a laser range finder mapped the topography of the lunar surface as the module orbited the Moon.

The records on film cassettes had to be retrieved on the return voyage, which could be done only by spacewalking from the command module to the scientific instrument bay. This was carried out in the weightless, vacuum conditions of space where the temperature in the shade is —270C. The astronaut was 100,000 miles from home at the time.

The splashdown of Apollo 17 came as a welcome climax. Recovery of a manned capsule is an expensive operation involving the deployment of substantial numbers of ships at strategic points on the oceans, but the Soviet technique of recovery on land is intrinsically more dangerous. The Space Shuttle, however, can be controlled in the same manner as a conventional aircraft, and can routinely permit safe and comfortable landings.

The end of the Apollo 17 mission on the Moon was recorded by a color television camera and simultaneously relayed to millions of viewers on Earth. In the later phases of the Apollo program considerable effort went into the provision of good quality television equipment so that the general public could follow, or be entertained by, the mission as it progressed. The lift-off of Apollo 17 on December 13, 1972, marked the end of a heroic effort, terminated because of its formidable expense once it had been demonstrated to work.

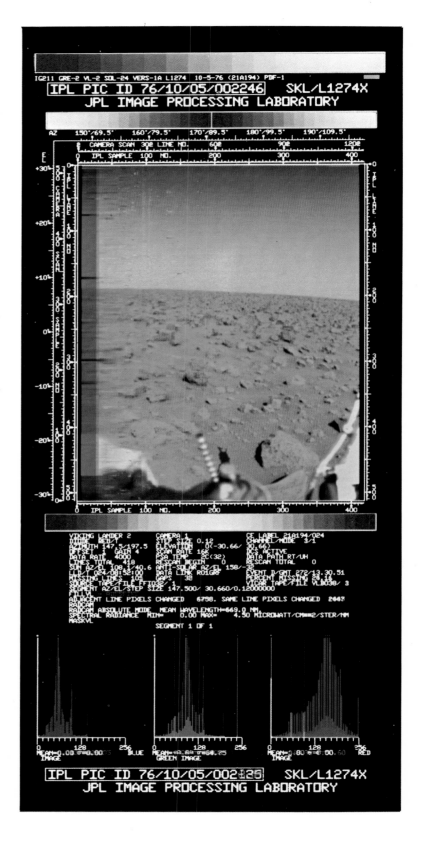

The red deserts and pink skies of Mars were photographed by Itek electro-optical cameras aboard the two Viking landers dispatched in 1976. In the camera being adjusted before the launch the rectangular scan mirror appears near the top. Each photograph was beamed back to Earth by way of a radio link on a Viking orbiter craft above Mars as a series of three monochromatic images. These were recombined at the image processing section of the Jet Propulsion Laboratory in Pasadena, California. Further computer enhancement enabled background interference and systematic error (such as the red band at the left of the frame) to be removed. The cameras, which worked for years after the landing on Mars on July 20, 1976, secured panoramic and close-up views of the surface soil and documented a variety of changes during the Martian seasons, particularly the redistribution of soil and windborne dust. The Viking landers were also fitted with equipment to measure temperature, pressure, wind speed, and atmospheric composition. Other apparatus was designed to conduct biological experiments and to search for signs of life.

NASA capped the highly successful Viking missions to Mars with long-distance visits to the outer planets, using the Voyager spacecraft. Two of them, weighing 1700 pounds, were launched in late 1977 and they reached Jupiter in 1979. Arrival at Saturn is scheduled for late 1980 and early 1982. The Voyagers are designed to take color photographs of superb quality quickly and reliably as the spacecraft skims past the planet and its satellites, and speeds further through the solar system. Consequently, all the photography has to be done at a single pass. Threading a fine line between the planet Jupiter and five of its 13 satellites, Voyager 1 put on an awesome nonstop show in February and March, 1979, as 15,000 images were sent back some 500 million miles of space to the ground station in California. This craft discovered a ring around Jupiter, like Saturn's but much fainter; volcanoes in action on the moon Io; and giant cracks in another moon, Europa.

A Voyager spacecraft is launched atop a Titan-Centaur rocket to begin its long journey to the two largest planets, Jupiter and Saturn. The liftoff thrust came from twin solid-propellant boosters, each 85 feet tall and 10 feet in diameter, built by United Technologies' Chemical Systems Division. Once in space, the Voyager was inserted into earth orbit and then sent on its way to the planets by two hydrogen-fueled rocket engines built by Pratt & Whitney Aircraft Group of United Technologies.

WHAT NEXT?

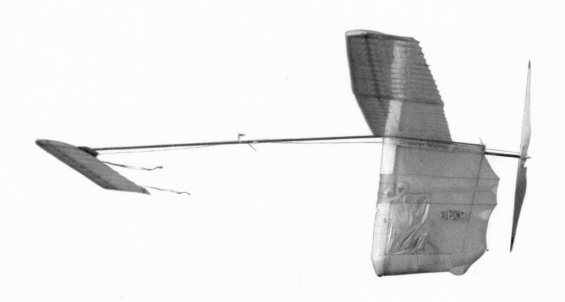

Clever beyond all dreams
the inventive craft that he has
which may drive him one time or another to well or ill.
When he honors the laws of the land and the gods' sworn right
high indeed is his city; but stateless the man
who dares to do what is shameful.

Sophocles: *Antigone.*

Overleaf: Bryan Allen made history on June 12, 1979 by pedaling Dr. Paul MacCready's Gossamer Albatross aircraft across the English Channel.

Technology is ethically neutral. There is nothing inherent in it that insures its use for good or evil. Neither mechanical genius nor the technocrat's knowledge will guarantee morality of action; for this, a higher wisdom is required.

The dilemma is as old as civilization and it was recognized by the thinkers of the most distant past. The fact is that technology develops cumulatively while ethics does not; technology is easy to impart while ethical wisdom is difficult to disseminate. The development of technology over the last centuries has been dramatic. But has it been proportionate to our ethical standards and our philosophical and social maturity? We would like to think Ortega y Gasset overstated the case when he wrote that modern man *possessed all the talents except the talent to make use of them.*

Indeed, technological development has helped man to solve a host of problems that plagued him over the centuries. Man, however, is confronted with new problems that technology has not found ways to cure. In this last quarter of the 20th century, amidst his unprecedented wealth and technology, modern man is feeling more insecure than ever; like the sorcerer's apprentice, he is haunted by the fear of having set in motion forces that he can no longer control.

The 1960s and '70s have witnessed a major cultural crisis. An increasing number of people reject the world we have created: nostalgic for a world we have lost, they delude themselves with idyllic and romantic visions of the past. But one can argue that what is needed is not a return to a more simple world and more primitive technologies. What is needed is to improve our own quality, to make us capable of putting our advanced technologies to good and profitable use. We must aspire to go forward. We do not need to go back.

NAME INDEX

The dates given for rulers or consorts are birth-accession-death

SUBJECT INDEX

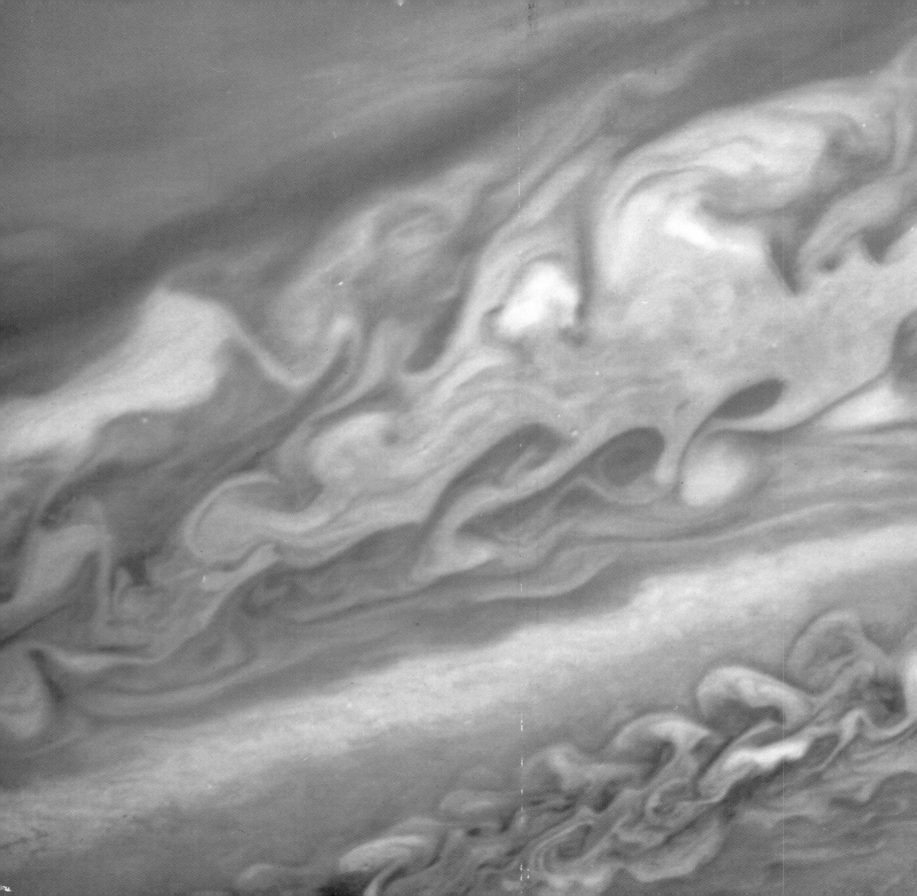